Vegetables!

Vegetables!

100 inspiring recipes
for every occasion

Pippa Cuthbert & Lindsay Cameron Wilson

Good Books

Intercourse, PA 17534
800/762-7171
www.GoodBooks.com

This edition copyright © 2008 Good Books, Intercourse, PA 17534

International Standard Book Number: 978-1-56148-621-2 (paperback edition)
International Standard Book Number: 978-1-56148-622-9 (comb-bound edition)
Library of Congress Catalog Card Number: 2007036750

Text and recipe copyright © 2008 Pippa Cuthbert and Lindsay Cameron Wilson
Photographs copyright © 2008 New Holland Publishers (UK) Ltd
Copyright © 2008 New Holland Publishers (UK) Ltd

Library of Congress Cataloging-in-Publication Data

Cuthbert, Pippa.
 Vegetables! : 100 inspiring recipes for every occasion / Pippa Cuthbert & Lindsay Cameron Wilson.
     p. cm.
 Includes index.
 ISBN 978-1-56148-621-2 (pbk. : alk. paper)
 ISBN 978-1-56148-622-9 (comb-bound : alk. paper)
 1. Cookery (Vegetables) 2. Vegetables.
 I. Wilson, Lindsay Cameron. II. Title.
 TX801.C875 2008
 641.6'51--dc22
                              2007036750

Senior Editor: Clare Sayer
Copy-Editor: Clare Hubbard
Design: Paul Wright
Photography: Stuart West
Food Styling: Pippa Cuthbert and Lindsay Cameron Wilson
Production: Hema Gohil
Editorial Direction: Rosemary Wilkinson

Reproduction by Pica Digital Pte Ltd, Singapore
Printed and bound in China by C&C Offset Printing Co

Note

The authors and publishers have made every effort to ensure that all instructions given in this book are safe and accurate, but they cannot accept liability for any resulting injury or loss or damage to either property or person, whether direct or consequential and howsoever arising.

Both metric and imperial measures are given in the recipes – follow either set of measures but not a mixture of both as they are not interchangeable.

All herbs used are fresh unless otherwise stated.

Oven temperatures are given for conventional ovens. If using a fan oven, reduce the oven temperature by 70°F (20°C).

Due to the slight risk of salmonella, children, the elderly and pregnant women should avoid recipes with lightly cooked or raw eggs.

Contents

Introduction

Pippa and I are from different ends of the earth – we drive on opposite sides of the road; water swirls down the drain in opposing directions; the evening sun sets in the east in Auckland, in Halifax you'll find it in the west. But there are some similarities. Asparagus ushers in spring, tomatoes flourish in summer and mushrooms abound in autumn. The seasons may be opposite – my spring is her autumn, her winter is my summer – but no matter. We both live by the rhythm of the garden.

Pippa and I both moved to London in 2001, when the trend towards eating seasonal, local foods was slowly gaining momentum. Cheryl Cohen of London Farmers' Markets says she can't pinpoint exactly when the popularity began. "I would say there has been a gradual awareness and awakening from many sides and angles – from food writers, food growers, people fed up with vegetables grown for appearance and sold by supermarkets at a premium rate, and an awareness that there is choice. It's like Chinese whispers: word of mouth that an alternative way of buying does exist."

The closest thing either of us had to vegetable gardens were tiny window-boxes filled with fresh herbs, so instead we practiced an alternative way of buying. On Saturday mornings we would join the throngs of tourists and locals alike at Borough Market. Sometimes it took a trip to the market to remind Pippa what was in season; it's hard to keep it straight when you spend time in two hemispheres. A few hours and several coffees later, we'd head back to our respective flats with our bounty. And clearly we weren't the only ones. Borough Market was last year's winner of the Visit London award for the most popular London experience. As Matthew Fort recently wrote in *The Guardian*, Borough Market has now "acquired iconic status, not just in London but also outside it. It is one of the great symbols of the revival of our food culture. Along with farmers' markets, it is proof that there is retail life outside supermarkets, that traditional shopping patterns are not entirely moribund, that consumer power counts for something and, above all, that there are people who care enough about their food to spend above the norm for quality."

When I moved back to Canada in the summer of 2004, I traded the Borough Market for the Halifax Farmers' Market. It was an easy switch. After three years, the same producers were still there – plus many more – chatting, smiling and selling all that Nova Scotia had to offer. And who knew how diverse the offerings could be? In the time that I was gone, there was a discovery, or rediscovery, of the seasons and the variety of produce, both old and new. I discovered local, organic black and green radishes, parsley root, rainbow chard, patty pans (a type of squash), purple carrots, golden beets, and white sprouting broccoli. I also ran into local chefs armed with palettes filled with whatever was good that week. They had a lovely symbiotic relationship with the farmers; they pushed each other to grow, or alternatively, to cook, something new. That is the beauty of growing organically on a small scale. When the crop is already diverse, why not add an experimental crop to the rotation?

Experimenting is what cooking is all about. There will always be a need to buy imported food – we Canadians might die of scurvy if we ate locally through the winter. And what would New Zealanders do without produce from the Pacific Islands? However, there is always an opportunity to augment with local produce. Buying locally supports farmers and cuts down on transport emissions – and you might discover varieties you haven't tasted since childhood. Local produce will vary, naturally, from place to place. Substitute ingredients in recipes with whatever catches your eye at the farmers' market. We've provided you with a culinary road-map, but veer off the beaten path. Let your local garden dictate your culinary rhythm.

Nutrition

Vegetables are an important source of many of the nutrients our bodies require for proper functioning. Whether sticking strictly to a vegetarian diet or not, it is important to understand where certain nutrients can be obtained in order to sustain a healthy lifestyle.

You may think it important to strictly define what a vegetable is, but the barriers have become blurred so we will keep it simple by leaving strict anatomical names and classifications aside. Despite the fact that we call and treat them as vegetables, green beans, peppers, pea pods and eggplants are all in fact anatomical fruits. You will, however, find these "vegetables" featured in our recipes, along with lentils and legumes. Think of this book more as a collection of well-balanced, meat-free recipes that will be suitable as a part of any vegetarian diet.

Nutritionists and medical experts universally recommend that we eat at least five servings of fruits and vegetables a day (on average

one serving = 3¼ ounces/ 80 grams). Fruits and vegetables should make up about a third of the food we consume daily. Recent studies suggest that adhering to a predominantly plant-based diet may decrease the risk of heart disease and certain types of cancer. Fruits and vegetables are low in fat and high in fiber as well as many vitamins and minerals. A diet based around vegetables, grains and legumes will also help maintain weight control as these foods are bulky and filling, yet contain virtually no fat. But as always, the rule is "everything in moderation."

VITAMINS

Vitamins are the most important nutritional contribution that vegetables make to our diet. There are two types of vitamins; fat-soluble vitamins and water-soluble vitamins.

Fat-soluble vitamins – vitamins A, D, E, and K – are found mainly in fatty foods such as animal fats, dairy products and vegetable oils, as well as in vegetables. These vitamins are essential for everyday functioning; however, our bodies have the ability to store these vitamins in the liver and fatty tissues until required. If consumed in excess of our needs, these vitamins can reach toxic levels.

Water-soluble vitamins – vitamins B6, B12, C, bioton, folic acid, niacin, pantothenic acid, riboflavin, thiamin – are found predominantly in fruits, vegetables and grains. Water-soluble vitamins are not stored in the body so need to be consumed regularly. These vitamins are not harmful if consumed in excess as any extra will be passed through the body.

Vitamin C:

Found in dark green, leafy vegetables, peppers, tomatoes and potatoes. Vitamin C – the most prominent of all vitamins in vegetables – is manufactured from the sugars that are supplied by the leaves as a result of photosynthesis. So in simple terms, the more sunshine and light your vegetables get, the more sugars they will produce and consequently the more plentiful the vitamin C. A useful guide is to look at the coloration of the leaves; a darker green will indicate higher levels of vitamin C. For example, the lighter inner leaves of a cabbage will often contain a fraction of the vitamin C of the outer darker leaves.

Benefits: Essential for the absorption of iron and the maintenance of healthy teeth and gums. Also aids in strengthening resistance against infection and certain diseases. Promotes cell renewal, aids in forming connective tissues and promotes healthy capillaries.

Vitamin A:
Found in dark green, leafy vegetables such as spinach and broccoli, and in orange vegetables such as carrots, pumpkins and sweet potatoes.

Benefits: Essential for maintaining healthy eyes and night-vision. Also promotes the growth and development of cells essential for healthy skin and bones.

Vitamin B2 (riboflavin):
Found in dark green, leafy vegetables, asparagus and broccoli.

Benefits: Essential for converting food into energy. It also aids normal growth of body tissues, promoting healthy skin and supporting vision.

Vitamin B6:
Found in green, leafy vegetables and legumes.

Benefits: Helps the formation of red blood cells and antibodies. Also helps the body build amino acids and proteins that are needed for proper growth and developmnent.

Vitamin E:
Found in green leafy vegetables, grains and soybeans.

Benefits: An important antioxidant that protects cell walls from oxidizing and thus ageing.

Folic acid (folate):
Found in dark green, leafy vegetables, cabbage and cauliflower.

Benefits: Essential for normal tissue growth and keeping cells healthy. Very important for women during pregnancy as it helps prevent defects of the spine in the unborn baby.

MINERALS
Minerals are required by our bodies in small amounts in order for us to function properly. The form that they are found in food is the form that our bodies require. Minerals have three main functions – building strong bones and teeth, controlling body fluids and turning the food we consume into energy. The major minerals the body needs (those required in larger amounts are) are calcium, phosphorous, magnesium, sodium and potassium. Trace minerals (those needed in smaller quantities) are iron, zinc, copper, selenium and iodine. Below is a list of the minerals and some of the vegetables from which you can obtain them.

Calcium:
Found in dark green, leafy vegetables such as spinach and watercress.

Benefits: Essential for healthy bones and teeth.

Phosphorous:
Found in cabbage, potatoes, peas and beans.

Benefits: Promotes healthy cells, bones and teeth.

Magnesium:
Found in dark green, leafy vegetables, such as cabbage and broccoli.

Benefits: Helps the body to use energy and aids effective muscle function.

Sodium:
Naturally occurs in almost all fresh, whole vegetables.

Benefits: Regulates the body's water content and helps the nerves to function effectively.

Potassium:
Found in dried peas and beans.

Benefits: Helps cells and body fluids to function properly.

Iron:
Although meat is the best source of iron it can also be found in a variety of beans and vegetables. Green vegetables such as broccoli, spinach, okra and watercress contain significant amounts of iron. Our bodies can more easily absorb the iron from these vegetables if consumed in conjunction with vitamin C, such as a glass of orange juice. Tea and coffee have the opposite effect and make the absorption of iron more difficult.

Benefits: Essential for the formation of red blood cells.

Zinc:
Most vegetables contain a certain quantity of zinc, but peas, lima beans and potatoes are vegetables that contain a significant amount.

Benefits: Helps the body reach sexual maturity and helps the body repair damaged tissue.

Copper:
Found in most vegetables, but particularly in lima beans.

Benefits: Helps the body to use iron efficiently.

Selenium:
Found in lima beans, peas, kale and sweet potatoes.

Benefits: Keeps the body's cells healthy.

Iodine:
All vegetables grown in iodine-rich soil will contain iodine.

Benefits: Aids the body's production of thyroid hormones, which control the body's metabolism.

PROTEIN
Protein is essential in our diets as it makes up almost all of the basic tissues that run our bodies on a day-to-day basis. Muscles, organs, blood cells, nails, hair and even our teeth and bones are all protein. These tissues are continuously being worn down and replaced, and the protein from our diets is used to replace them.

Whole grains, legumes, soy products and nuts and seeds are primary protein sources in a plant-based diet; however vegetables do contribute to our daily requirements also. Dairy products and eggs are another great source of protein for those sticking to a meat-free diet, though it is advisable to choose low-fat dairy products if consuming large quantities.

The question is, how much protein do our bodies require? Recommended daily allowances (RDA) vary internationally. On the whole, women generally need less protein than men. However, women who are pregnant or breastfeeding have greater requirements. Always remember that too much protein in our diets is converted to fat and can also

put excess strain on the kidneys and liver.

When consuming a plant-based diet, it is important to understand the difference between "complete" and "incomplete" proteins. Complete proteins contain all the essential amino acids that our bodies require. All animal foods, i.e. meats, eggs and milk products are complete proteins, because animals are genetically built in a similar way to humans. Plant proteins, however, are generally incomplete due to the marked difference in genetic make-up between plants and humans. There are some exceptions to these rules. For example, wheat germ and soy proteins are nearly complete, while gelatin, which is extracted from animal skin and bone, is so incomplete it cannot sustain life at all.

Human manipulation can make a difference; other food proteins can complement plant proteins to provide us with our complete protein requirements. The marvelous human body enables us to store usable amino acids in our bodies for up to 12 hours, enabling time for complementary amino acids to be consumed. By eating a variety of foods throughout the day, you can ensure that you are meeting all your requirements.

Cooking vegetables

The moment a vegetable is plunged into boiling water or exposed to a hot oven, its make-up is altered in many ways. Cooking vegetables affects flavor, texture, color and nutritional value.

Flavor is more often than not intensified when vegetables are cooked, as high temperatures will make the aromatic molecules more volatile and consequently easier to detect. However, if a vegetable is overcooked, the aromatic molecules may be destroyed, giving eventual loss of flavor.

Texture is probably the attribute affected most when heat is applied. The texture of vegetables is determined by both cell structure and the amount of water in the tissues. When heat is applied, the cell structure weakens and the water is extracted. In most vegetables, particularly starchy ones, we require the cell structure or starch granules to be altered before the vegetable is desirable to eat. Other vegetables, such as green

ones which do not necessarily require cooking, become more digestible when exposed to a short amount of cooking. The best way to judge whether a vegetable is cooked comes down to experience and personal taste. These days we seem to prefer our vegetables tender but still firm or *al dente* as the Italians say. Only a generation ago, the preference was for softer, mushier vegetables.

The color of green, leafy vegetables and stem vegetables undergoes significant changes when heat is applied. Strangely enough, the color change has very little to do with the color pigments involved. It is the result of a sudden expansion and escape of the gases that are trapped between the cells. The result is an intensely bright coloration which then dulls the longer the vegetable is cooked. This change takes place very rapidly after immersion in boiling water. The less cooking the vegetable requires, the brighter the coloration will be. One exception is the carotenoid family which include carrots and tomatoes. This group of vegetables will lose very little, if any, color during cooking. If you

do not require your vegetables hot, then plunge them into ice-cold water straight after cooking to retain maximum color.

Although cooking can affect the nutritional content of vegetables and destroy some vitamins, losses are not huge (usually only about 25 percent). To minimize nutritional loss, try shorter cooking times, steaming rather than boiling and using fresh, seasonal produce. Fat-soluble vitamins, such as vitamins A, D, E and K, are not affected by heat and therefore remain in the food during cooking. Water-soluble vitamins, including almost all other vitamins, are lost in boiling water. Some are also sensitive to air, heat and light.

Spoilage and storage

Vegetables should always be eaten as quickly as possible after picking. Ideally everyone would be self-sufficient, with a vegetable garden in the backyard, but we understand that this is often impractical. The next best option is that you are able to purchase vegetables from local farmers' markets. But again, not always practical or possible. Inevitably there is almost always a lapse in time between harvesting food and using it. Unfortunately vegetables begin to deteriorate in sweetness, flavor and texture from the moment they have been picked. The reason is that the plant cells try to continue their normal processes even though they no longer have their own source of food and water. Corn and peas lose up to 40 percent of their sugar in 6 hours at room temperature. Other vegetables, such as asparagus, begin to use their sugars to produce tough, indigestible fibers. We try to avoid using vegetables that have gone through further treatments such as freezing and canning, as our emphasis is always on using

"fresh" ingredients, but there are some pantry staples that we can't do without. Here are a few guidelines that may be useful:

- refrigerate vegetables immediately after purchase to slow down the metabolic activity in the cells. Refrigeration also slows down the activity of spoilage microbes if present.

- one piece of moldy or rotten produce will spoil anything it is in contact with. Always remove the offending item immediately!

- avoid putting vegetables under physical stress. Don't squash them into your fridge but instead arrange them carefully.

- remove dirt or soil from less delicate vegetables as soon as you buy them as these contain microbes which may increase spoilage.

- wash stem and leaf vegetables in water and keep moist to prevent wilting.

- plants are still living and require oxygen to respire. Avoid storing vegetables in airtight bags or containers. Paper bags are not only more environmentally-friendly, but will also allow your vegetables to breathe.

Main courses

Always the bridesmaid, never the bride. This has long since been the status of vegetables. They are so often the overcooked afterthought to the "meat and two veg" equation. They have been there, the silent wallflowers, while something meaty is always in their way, stealing the spotlight.

But times are changing. Jeffrey Steingarten, the food columnist for *Vogue*, recently witnessed this shift. He was granted two weeks with private chef, Paul Liebrandt. What Steingarten noticed most about his time with Liebrandt was the subtle changes he made to the way they ate. "For one thing," writes Steingarten, "every meal included several vegetables, each deliciously cooked for its own gastronomic value and not simply because one is supposed to eat one's vegetables."

We don't expect you to have Liebrandt's commercial stove equipped with immersion circulators to carry out *sous-vide* cooking. We won't ask you to turn asparagus juice into foam, nor will we expect tiny vegetable portions to appease your appetite. No, this is hearty, honest cooking designed to please any palate. Finally, the vegetable is getting married.

Grown-up mac and five cheese

Nostalgic

All right, the idea for this dish came when no one ate cheese at my wine and cheese party. Adding lovely cheese is, admittedly, a decadent way to enjoy macaroni. But who can argue when the king of comfort foods is elevated to an even higher status?

Serves 6

1lb 2oz (500g) **macaroni**
1 Tbsp + 1 tsp **salt**
1 stick (120g) **butter**
⅔ cup (75g) **all-purpose flour**
1 small **onion**, peeled and finely chopped
2 tsp **Dijon mustard**
½ tsp **Worcestershire sauce**
¼ tsp **freshly cracked pepper**
4½ cups (1L) **milk**
7oz (200g) **goat's cheese**, crumbled
7oz (200g) **aged Cheddar**, grated
7oz (200g) **Camembert**, roughly chopped
4oz (100g) **blue cheese**, crumbled
4oz (100g) **sundried tomatoes**, roughly chopped
3oz (75g) **Parmesan cheese**

Grease a large ovenproof casserole dish and set aside. Preheat the oven to 350°F (180°C).

Bring a large stockpot full of water to a boil. Add the macaroni and 1 Tbsp salt. Stir well and boil gently according to the directions on the packet. Drain well and set aside.

While the macaroni is cooking, melt the butter in a large saucepan over medium heat. Add the flour and stir with a wooden spoon until mixture is combined, about 1–2 minutes. Add the onion, salt, mustard, Worcestershire sauce and pepper; stir well to combine. Add the milk, slowly, stirring constantly, until the mixture is smooth and thickened. Be patient, this will take 8–10 minutes. Add the cheeses (except for the Parmesan), cooked macaroni and sundried tomatoes. Stir well.

Transfer to the prepared casserole dish and top with grated Parmesan cheese. Bake for 40–45 minutes, until the topping is golden.

Spinach, roasted pepper and ricotta lasagne

Comforting

The sweet, succulent roasted peppers add a delightful touch to this vegetarian lasagne dish.

Serves 6

For the tomato sauce:
1 Tbsp **olive oil**
1 **onion, peeled and finely chopped**
2 **garlic cloves, peeled and finely** chopped
1 Tbsp **tomato paste**
14-fl oz (400-ml) **can chopped tomatoes**
1 tsp **sugar**
Large handful **fresh basil leaves,** roughly torn

7½ cups (350g) **fresh or frozen spinach**
2¼ cups (500g) **ricotta**
1 **egg**
Pinch **nutmeg**
Salt and freshly ground black pepper
1 **quantity Béchamel sauce (see page 158) with** ¼ cup (50g) **grated Parmesan cheese added**
3 **roasted red peppers (see page 44), peeled, deseeded and torn into large strips**
9oz (250g) **lasagne sheets**
Extra grated Parmesan cheese
Salt and freshly ground black pepper

To make the tomato sauce, heat the oil in a large heavy-based saucepan. Add

the onion and sauté until cooked and translucent but not browned. Stir in the garlic and sauté for a further minute, then add the tomato paste. Add the tomatoes, sugar and basil leaves and bring to a boil. Reduce the heat and simmer for about 20–25 minutes.

Now prepare the spinach. Bring a large pan of water to a boil, add the spinach and cook until just wilted. Drain and allow to cool slightly before squeezing out any excess water. Chop the spinach finely and transfer to a large bowl. Add the ricotta, egg, nutmeg and season generously with salt and pepper. Set aside.

Preheat the oven to 400°F (200°C). Take a 9–10-cup (2–2.25-L) lasagne pan and put a third of the Béchamel sauce in the pan, top with a layer of lasagne sheets, half the tomato sauce, half the ricotta mix, lasagne sheets, a third of the Béchamel sauce, two-thirds of the roasted pepper slices, lasagne sheets, remaining tomato sauce, remaining ricotta mix, lasagne sheets, remaining Béchamel sauce. Top with the remaining pepper slices, an extra sprinkling of grated Parmesan cheese and freshly ground black pepper. Bake in the oven for 25–30 minutes. Leave to rest for 10 minutes before serving.

Linguini with sweet onions and Swiss chard

Linguini with sweet onions and Swiss chard

Sweet

Also known as lead beet, sea kale beet, white beet or spinach beet, the stalks of Swiss chard, a wintry green, range from a pale celadon color to a vivid magenta. This dish is especially pretty made with rainbow chard, the colorful cousin to Swiss chard, but any chard will do. Coated in meltingly sweet onions and a sprinkling of goat's cheese is a warming, colorful way to go.

Serves 4

2 Tbsp **butter**
2 Tbsp **olive oil**
2 large **onions, peeled and finely sliced**
1lb 2oz (500g) **linguini**
1 Tbsp + ½ tsp **salt**
1 bunch (about 6¼ cups/300g) **Swiss or rainbow chard**
½ tsp **freshly ground black pepper**
½ cup (125ml) **dry white wine**
4oz (100g) **goat's cheese, crumbled**

Heat the butter and olive oil in a pan over medium-high heat. Add the sliced onions and stir to coat the onions in the butter and oil. Cover the pan, reduce the heat to low, then leave the onions to slowly cook for 1 hour or until very soft. Uncover and continue cooking the onions until dark golden, about a further 25 minutes.

While the onions are cooking, prepare the chard. Cut 1½in (3cm) from the bottom of the stalks and discard. Slice the chard crosswise into ¾-in (2-cm) wide slices.

Meanwhile, bring a large pan of water to a boil. Put the linguini in the pan, along with 1 Tbsp of salt and cook according to directions on package.

Season the onions with salt and pepper, increase the heat, then add the wine. When the bubbles subside, add the chard. Stir well, turn the heat to medium-low, cover and stew until chard is wilted and tender, about 10 minutes.

Drain the pasta and toss with onion and chard mixture. Divide among four bowls and top with crumbled goat's cheese and add more salt and pepper to taste. Serve immediately.

■ *The onions can be made several hours in advance and reheated when needed.*

Baked ricotta with sundried tomatoes and baby salad leaves

Fresh

This is a perfect light summer lunch and is so simple to prepare. The baked ricotta can be made the day before and refrigerated. Just bring it to room temperature before serving.

Serves 4–6

For the baked ricotta:
2¼ cups (500g) **ricotta**
2 **eggs,** lightly beaten
Finely grated **zest of** 1 **lemon**
1 tsp **cracked black pepper**
2 **garlic cloves,** peeled and crushed
2 Tbsp finely chopped **fresh marjoram, oregano or thyme**
Salt to taste
2 Tbsp **pumpkin seeds**
1 Tbsp **sesame seeds**

2 Tbsp **extra-virgin olive oil**
1 Tbsp **balsamic vinegar**
Salt and freshly ground black pepper
5 cups (200g) **mixed baby salad leaves or mixed chard leaves**
4oz (100g) **sundried or semidried tomatoes**
2 Tbsp **pumpkin seeds,** toasted

First, make the baked ricotta. Preheat the oven to 350°F (180°C). Grease the base and sides of a 7-in (18-cm) springform cake pan and line the base with parchment paper. Mix the ricotta with all of the remaining ingredients, except the seeds, adding salt to taste. Spoon the mixture into the prepared pan, sprinkle over the seeds and bake for 40–45 minutes or until golden brown and set. Remove from the oven and leave to cool in the pan while you prepare the salad.

Combine the oil and balsamic vinegar and season generously with salt and pepper. Toss the dressing through the salad leaves and tomatoes.

When the ricotta has cooled slightly, remove it from the pan and cut into wedges to serve. Serve a wedge of ricotta on each plate topped with a handful of salad and a sprinkling of toasted pumpkin seeds.

■ *Pumpkin seeds can be toasted in a matter of minutes: simply place a small heavy-based frying pan over a medium heat and when hot, add the seeds. Shake gently for a minute until they turn golden and take care not to let them burn!*

Sweet potato and ricotta gnocchi with pecans

Festive

A sophisticated, grown-up take on the classic Thanksgiving sweet potato and pecan side dish. Making your own gnocchi is a much easier task than you might think. Experiment with other sauces to suit your taste.

Serves 6

For the gnocchi:
5 medium (800g) **sweet potatoes**
1¾ cups (400g) **ricotta, drained in a sieve for 2 hours**
3oz (75g) **Parmesan cheese, grated, plus extra to garnish**
2 tsp + 2 Tbsp **salt**
½ tsp **freshly ground nutmeg**
3 cups (350g) **all-purpose flour**

For the sauce:
1 stick (100g) **butter**
2 Tbsp **maple syrup**
¼ tsp **crushed red pepper**
½ cup (50g) **pecans, toasted and roughly chopped**
3 cups (50g) **fresh flat-leaf parsley, roughly chopped**
Sea salt and freshly ground black pepper to taste

Line a baking tray with parchment or wax paper and set aside.

Place the unpeeled sweet potatoes in a large saucepan, cover with cold water and boil until tender. Drain in a colander. When cool enough to handle, peel the potatoes and place in a large bowl. Add the ricotta, 3oz (75g)

Parmesan cheese, 2 tsp salt and nutmeg. Mash until smooth. Slowly add flour, until the dough is soft.

Turn the dough out onto a floured surface and divide it into six equal pieces. Roll each piece into a 20-in (50-cm) long, 1-in (2.5-cm) diameter rope, using your palms. Sprinkle the dough with flour if it becomes sticky. Cut each rope into 20 pieces. Roll each piece under the prongs of a fork to indent. Transfer to the baking tray.

Bring a large pan of water to a boil; add 2 Tbsp salt and bring back to a boil. Working in batches, boil the gnocchi until tender, 5–6 minutes. Transfer the cooked gnocchi to a clean baking tray and cool completely.

Just before serving, make the sauce. Heat the butter in a large frying pan over medium heat. When the bubbles subside and the butter is dark golden (not brown), add the maple syrup and crushed red pepper. Stir to coat, then turn up the heat to medium-high and add the gnocchi and sauté until crispy. Transfer to serving bowls and top with the pecans, parsley, grated Parmesan cheese and salt and pepper to taste.

Barley risotto with roasted butternut squash and goat's cheese

Barley risotto with roasted butternut squash and goat's cheese

Nutty

I love the nutty texture and flavor of the barley in this dish. It is served in the usual Italian way, *al dente*, but you can cook it for longer if preferred.

Serves 4

1 **butternut squash or pumpkin** (about 1¾–2¼lb/800g–1kg in weight)
1 **red onion,** peeled and cut into thin wedges
2 Tbsp **olive oil**
Salt
1 Tbsp **butter**
3 **garlic cloves,** peeled and crushed
2 cups (250g) **pearl barley,** rinsed
1 cup (250ml) **white wine**
4½ cups (1L) **hot vegetable stock**
Small handful **flat-leaf parsley,** chopped
3oz (75g) **goat's cheese,** cut into small cubes
2½ cups (100g) **watercress or arugula,** tough stalks removed
Freshly ground black pepper

Preheat the oven to 400°F (200°C). Peel and cut the squash into approximately ¾-in (2-cm) cubes. Put the squash and red onion into a roasting pan and toss with 1 Tbsp of the oil. Season generously with salt and cook for 20–25 minutes, stirring once or twice until golden and cooked.

While the squash is cooking, prepare the remaining ingredients. Heat the remaining oil and the butter in a large, heavy-based saucepan and stir in the garlic and barley. Toss to coat in the oil and butter, and stir for a couple of minutes. Add the wine, bring to a boil, and then simmer until the wine evaporates. Add a ladleful of hot stock. Each time the liquid evaporates add another ladleful of stock until it has all been added. Reduce the heat and simmer, stirring occasionally for 30–40 minutes. The barley should become tender but the mixture still remain slightly soupy. Stir in the parsley, goat's cheese and half of the watercress, along with the squash and red onion.

Season to taste and serve with the remaining watercress or arugula on top.

Broccoli, chili and almond penne

Textured

This dish will appeal to young and old alike. It is a great way to disguise broccoli for the kids, but you may want to leave out or decrease the chili if serving it to children, and maybe try adding a little lemon zest instead.

Serves 4

1½ cups (350g) **broccoli florets**
2 Tbsp **olive oil**
1 **onion,** peeled and finely diced
¼–½ tsp **crushed red pepper**
2 **garlic cloves,** peeled and crushed
½ cup (50g) **slivered almonds or pine nuts**
⅔ cup (150g) **mascarpone**
Salt and freshly ground black pepper
14oz (400g) **penne pasta**
Grated Parmesan cheese to serve (optional)

Bring a large saucepan of water to a boil. Add the broccoli and cook until tender but still holding its shape. Remove from the heat and drain.

Heat the oil in a large, nonstick frying pan. Sauté the onion for 3–4 minutes or until soft and translucent. Add the crushed red pepper and garlic and sauté for a further minute. Then add the almonds and cook, stirring, until the almonds turn lightly golden. Remove from the heat and add half of the onion mixture to the broccoli.

Using a hand blender or food processor, pulse until most of the broccoli is puréed, but it is fine to have a few smallish chunks. Add the puréed broccoli mixture back into the frying pan with the remaining onion mixture and add the mascarpone. Stir well to combine and season to taste.

Cook the pasta according to directions on the packet, drain, and then stir through the sauce. Serve immediately with a sprinkling of Parmesan cheese if liked.

Caramelized red onion and sweet potato tart

Warming

For this recipe I like to use kumara, a unique variety of sweet potato grown in New Zealand. But any sweet potato variety will do.

Serves 4–6

3 medium (400g) **kumara or sweet potatoes,** peeled and cut in half or into large chunks
1 Tbsp **butter**
1 Tbsp **extra-virgin olive oil**
3 medium (600g) **red onions,** peeled and thinly sliced
Salt and freshly ground black pepper
2 Tbsp **sugar**
2–3 **sprigs thyme,** leaves removed
2 Tbsp **sherry or balsamic vinegar**
½ cup (100ml) **hot vegetable stock or boiling water**
9oz (250g) **puff pastry**
4oz (100g) **dolcelatte or goat's cheese**

Preheat the oven to 400ºF (200ºC). Bring a pan of water to a boil, add the kumara and cook for about 8–10 minutes or until cooked but still firm. Drain and set aside to cool slightly.

In a 9½-in (24-cm) diameter nonstick, ovenproof frying pan, heat the butter and oil. Add the red onion and sweat over medium heat until soft but not browned. Season generously with salt and pepper, add the sugar and thyme leaves and increase the heat. Stir in the vinegar and all the stock and cook, stirring occasionally, until the liquid has reduced and the onions are sticky and caramelized, about 20 minutes. Remove from the heat.

Cut the slightly cooled kumara into cubes about ¾-in (2-cm) square and mix into the caramelized onions.

Roll the pastry out into a circle slightly larger than the frying pan and prick with a fork to make holes for the steam to escape. Cover the onion mixture with the pastry, tucking the edges well in around the side of the pan. Cook for about 35–40 minutes or until the pastry is crisp and golden brown.

Rest the tart for about 5–10 minutes before turning it out. Run a knife between the edges of the tart and the pan. Put a plate over the pan and turn upside down onto the plate. Crumble over the dolcelatte or goat's cheese and serve warm.

■ *We like to do a variation of this tarte tatin using potato, rosemary and caramelized shallots. Choose a potato that is slightly waxy and will hold its shape, such as a Desirée or new potato.*

Broiled Portobello mushrooms with celeriac mash

Warming

This is hearty, warming, Sunday lunch fare. The grilled Portobello mushrooms are also delicious tucked within a multigrain bun and topped with your favorite burger condiments.

Serves 4

4 large Portobello mushrooms, stems removed
½ cup (125ml) olive oil
¼ cup (75ml) red wine vinegar
2 Tbsp Dijon mustard
2 garlic cloves, peeled and thinly sliced
1 Tbsp each chopped fresh thyme, rosemary and/or basil leaves
½ tsp sea salt
¼ tsp freshly ground black pepper
10 cups (500g) baby spinach, washed
1 quantity of Celeriac mash (see page 85) to serve

Place the mushrooms in a shallow baking dish, or a plastic resealable food storage bag. Whisk the remaining ingredients (except the Celeriac mash and spinach) together in a small bowl. Drizzle over the mushrooms, turning the mushrooms well to coat. Leave to marinate for up to 3 hours in the refrigerator or 1 hour at room temperature, turning the mushrooms every 20 minutes or so.

Preheat the broiler. Place the mushrooms, underside up, on a baking tray. Spoon over any remaining marinade. Slide the tray under the heat and broil for 3–4 minutes, until cooked through and very juicy.

While the mushrooms are under the broiler, place a large frying pan over medium heat. Add the damp spinach and stir until wilted. Set aside.

Serve the mushrooms with Celeriac mash and wilted spinach, and top with any juices from the baking tray.

Broiled Portobello mushrooms with celeriac mash

Classic herb and chèvre omelette

Adaptable

An omelette is the perfect meal for any time of day – whether it be breakfast, lunch or dinner. There's something very comforting about its simplicity and ease.

Serves 1

3 eggs, preferably free range and organic
Salt and freshly ground black pepper
1 Tbsp (15g) **unsalted butter**
2 Tbsp chopped **mixed herbs** (e.g. chives, basil, parsley, mint)
1–2oz (25–50g) **chèvre**

Gently beat the eggs and season with salt and pepper.

In a 10–12-in (25–30-cm) nonstick frying pan, heat the butter until melted and foaming. Throw in the chopped herbs and stir briefly until vibrant green. Pour in the beaten eggs, tipping the pan so that the eggs cover the bottom of the pan. Crumble the chèvre over the omelette and cook, lifting the sides and swirling the pan until the top is very nearly cooked. Flip the sides in, folding the omelette into thirds, it should be lightly golden on the underneath. Transfer to a plate and serve immediately.

Braised eggplant, olives and prunes on basmati rice

Succulent

The Mediterranean meets the Middle East in this rich, sophisticated (and very easy!) dish. Thanks to the Turner family for the inspiration! Serve over basmati rice or the rice of your choice.

Serves 4–6

1 Tbsp plus ½ cup (125ml) **olive oil**
1 **large eggplant,** cut into 1¼-in (3-cm) cubes
1 **head garlic,** each clove peeled
4 Tbsp **dried oregano**
⅔ cup (150ml) **red wine vinegar**
1¼ cups (200g) **dried prunes**
1 cup (100g) **kalamata olives,** pitted
4 Tbsp **capers,** rinsed
6 **bay leaves**
½ tsp **sea salt**
¼ tsp **freshly ground black pepper**
1 cup (150g) **brown sugar**
1 cup (250ml) **white wine**
¾ cup (15g) **fresh flat-leaf parsley,** roughly chopped

Heat 1 Tbsp olive oil in a large frying pan over medium-high heat. Add the eggplant and fry, stirring occasionally, until golden brown, about 15 minutes. Remove from the heat and cool. Combine the cooked eggplant, garlic, oregano, red wine vinegar, olive oil, prunes, olives, capers, bay leaves, salt and pepper in an ovenproof casserole dish. Cover and marinate in the refrigerator overnight or for at least 8 hours.

Preheat the oven to 325°F (160°C) and take the casserole dish out of the fridge to bring to room temperature. Add the brown sugar, white wine and two-thirds of the parsley to the casserole and stir well. Bake for 45 minutes, or until the eggplant is very tender and the center is bubbling.

Serve over basmati rice and garnish with the remaining parsley.

Leek, fennel and basil tart

Subtle

These subtle flavors all amalgamate into complete bliss. Try using rice or dried beans to line your pastry shell if you don't have any baking beans.

Serves 6

13oz (375g) **shortcrust pastry or pie dough**
2 Tbsp **extra-virgin olive oil**
¼ stick (30g) **butter**
1 **onion,** peeled and finely chopped
1 **fennel bulb,** finely diced
12oz (350g) **leeks,** thinly sliced
1 **garlic clove,** peeled and crushed
⅔ cup (150g) **mascarpone**
¼ cup (75ml) **heavy cream**
4 **large eggs,** beaten
2oz (50g) **Parmesan cheese,** grated
Large handful **fresh basil leaves,** roughly torn
Salt and freshly ground black pepper

On a lightly-floured surface, roll out the pastry to ⅛in (3–4mm) thick and use to line a 10-in (28-cm) loose-bottomed tart pan. Chill the pastry for about 20 minutes. Preheat the oven to 400°F (200°C).

When the pastry has chilled, line it with baking paper and ceramic pie weights and bake for 15 minutes. Remove the baking paper and weights and return to the oven for a further 5 minutes or until lightly golden.

Meanwhile prepare the filling. In a large heavy-based saucepan, heat the oil and butter. Add the onion, fennel and leeks and sweat until cooked and tender but not browned, about 6–8 minutes. Stir in the garlic and cook for a further minute, then remove from the heat.

In a separate bowl combine the mascarpone, cream and eggs and lightly beat together. Add the cream mixture, Parmesan cheese and basil leaves to the leek mixture and stir to combine. Season generously with salt and pepper.

Reduce the oven temperature to 350°F (180°C). Pour the filling mixture into the warm pastry shell and bake in the oven for 25–35 minutes or until set and golden on top. Serve warm or at room temperature.

Leek, fennel and basil tart

Pan bagnat

Colorful

Using the key flavors of an antipasto platter, this is a vegetarian's picnic delight. Try using one large, hollowed out crusty loaf if preferred and any combination of fillings. Canned tuna, egg and capers are also favorite additions.

Serves 2–4

2 x 15¾-in (40-cm) **crusty baguettes**
2 Tbsp **extra-virgin olive oil**
1¼ cups (200g) **marinated artichokes**, halved
2 small (200g) **roasted red peppers**, peeled, cored, deseeded and cut into strips
12 **black olives**, pitted and halved
7oz (200g) **mozzarella**, sliced
4¾ cups (200g) **sundried tomatoes**, halved

Using a bread knife cut each baguette horizontally into two halves. Pull some of the bread from the center of each half to form a tunnel. Brush about 1 Tbsp of oil onto both sides of each baguette. Divide the remaining ingredients between the baguettes, layering and packing them in tightly.

Replace the top halves of the baguettes, then tightly wrap in clear food wrap. When ready to serve, cut the rolls into 2–4 pieces, remove the food wrap and serve immediately.

■ *These can be made up to two days in advance.*

Pan bagnat

Pumpkin ravioli with sage and pine-nut butter

Impressive

I learned how to make pasta in Italy when I was working on cooking courses a few years back. It's a lot easier than you would expect and definitely a great way to impress guests – believe me, I've done it!

Serves 4–6

Pumpkin filling:
1 (450g) **pumpkin or butternut squash,** peeled, deseeded and chopped
1 tsp **olive oil**
1¼oz (30g) **Parmesan cheese,** grated
Pinch freshly grated **nutmeg**
4oz (100g) **ricotta**
Salt and freshly ground black pepper

1 **quantity Egg pasta dough (see page 159)**

For the butter:
1 stick (100g) **butter**
12–18 **sage leaves**
½ cup (50g) **pine nuts,** lightly toasted

First prepare the filling. Preheat the oven to 400°F (200°C). Put the pumpkin in a roasting pan and drizzle with the oil. Toss to coat, then roast in the oven, stirring occasionally, for 25–30 minutes or until soft and tender. Remove from the oven and leave to cool slightly.

When cool enough to handle, mash the pumpkin using a fork. Add the Parmesan, nutmeg and ricotta and mix well until smooth. Season generously with salt and pepper. Set aside.

Roll out the pasta to very thin – you should be able to see your fingers through it. Using a 2–2¼-in (5–6-cm) round, fluted cutter, cut out 48–60 rounds, or as many as you can get out of the pasta. Using a teaspoon, put a small amount of the filling in the center of half of the pasta rounds. Using your finger or a pastry brush, brush a little water around the edge of the disc and place a second disc on top, pressing the edges to seal. Use a fork to press the edges for a more decorative finish. As you go, set the ravioli aside on a floured tray until all are done.

Bring a large pan of cold water to a boil, reduce the water to a simmer then gently poach the ravioli for 4–5 minutes. Drain well.

Meanwhile prepare the butter. Heat the butter gently over a low heat with the sage leaves. Stir in the pine nuts then pour over the ravioli and serve immediately.

Pumpkin ravioli with sage and pine-nut butter

Ratatouille

Provençal

This is one of the first local dishes I learned to make while studying in Provence, way back when. Every time these ingredients come together in my kitchen, I am taken back to our little kitchen in Aix, where little bowls filled with sprigs of fresh thyme sat ready for the taking. Ratatouille is an incredibly versatile dish – serve it warm or cold, over crusty bread or spooned over pasta. Add Pepita pesto (see page 153) before serving for an extra zing or use leftovers as a filling for Ratatouille strudel with basil and chèvre (see opposite). Ratatouille is even better if left for a day before eating.

Serves 6

1 large eggplant
1 tsp salt
2 Tbsp olive oil
3 garlic cloves, peeled and minced
3 onions, peeled
3 red, green and yellow peppers, cored and deseeded
3 zucchini or summer squash
5 tomatoes (see tip below)
5 sprigs fresh thyme
Sea salt and freshly ground black pepper to taste

Cut the eggplant into cubes, about ¾-in (2-cm) square. Place them in a colander and sprinkle with salt (this will remove the bitter liquids). Chop the onions, peppers, zucchini and tomatoes into cubes of a similar size to the eggplants, but keep them separate.

Rinse the eggplant and pat dry with paper towel. Heat the olive oil in a large saucepan over medium heat. Add the eggplant and sauté until golden. Remove the eggplant with a slotted spoon and set aside.

Add the onions to the pan and fry, stirring occasionally, until soft, about 10 minutes. Add the garlic and peppers and fry for 2 minutes. Add the zucchini and fry for a few more minutes. Add the tomatoes and thyme and cook for 10 minutes, stirring occasionally. Finally, return the eggplant to the pan and cook for a further 20–25 minutes, stirring occasionally. Season with salt and pepper to taste. Remove the thyme sprigs before serving.

■ *Use fresh tomatoes if they are sweet and succulent. If not, substitute organic, canned Italian tomatoes.*

Ratatouille strudel with basil and chèvre

Celebratory

This dish is the perfect vehicle for leftover ratatouille, caponata or any thick vegetable ragù you have at hand. However, this recipe is impressive enough not to be classified as a "leftover"; in fact, don't even mention that part.

Serves 4

5 **sheets filo pastry**
2 Tbsp **olive oil**
2 Tbsp **cold water**
1¼ cups (300ml) **Ratatouille (see opposite) or a thick, chunky vegetable sauce**
4 Tbsp **freshly-cooked couscous or quinoa**
4oz (100g) **chèvre**
½ cup (10g) **fresh basil leaves, torn**
2 Tbsp **pine nuts, toasted**
Green salad to serve

Preheat the oven to 400°F (200°C). Place a clean tea towel on a work-surface. Place one filo sheet on the towel. Whisk together olive oil and water in a small bowl. Brush over the filo sheet. Continue layering and brushing with remaining sheets.

Combine the Ratatouille and couscous in a bowl. Spoon the filling onto one long edge of the filo stack. Top with chèvre, fresh basil leaves and the pine nuts. Roll lengthwise, using the tea towel to guide the filo along.

Transfer to a baking tray. Cut through the top layer of the strudel, marking four equal-sized pieces. Bake for 40 minutes or until golden and crisp.

Serve immediately with a green salad.

Roasted pepper, olive and taleggio pizza

Vibrant

This is just a starting point to an endless array of pizza toppings. If you can't get hold of taleggio cheese, mozzarella works just as well.

Makes 2

For the pizza base:
2 tsp **dried granular yeast**
⅔ cup (150ml) lukewarm (boiled) water
2¼ cups (250g) **all-purpose flour (plus extra for kneading)**
½ tsp **salt**

For the topping:
2 **red peppers**
1 **yellow pepper**
1 tsp **extra-virgin olive oil**
4 Tbsp **Salsa verde (see page 164)**
½ cup (50g) **black olives**
4oz (100g) **taleggio (or mozzarella) cheese,** sliced
Freshly ground black pepper
Small handful fresh basil leaves

First, prepare the pizza bases. Sprinkle the yeast into ¼ cup (50ml) of the water and leave for 5–10 minutes. Add 1 Tbsp of flour and mix to a smooth paste, then stir in the remaining water. Cover and leave the yeast mixture for about 30 minutes or until it is bubbling and foamy.

Combine the flour and salt in a large bowl and make a well in the center. Pour in the yeast liquid. Using a wooden spoon, work the ingredients together by pulling the flour into the liquid. Use your hands to transfer the mixture to a lightly-floured surface. Knead the dough for 10 minutes until smooth and elastic. Form into a round loaf and leave to rise under a clean tea towel for about 1½ hours or until doubled in size.

Preheat the oven to 350°F (180°C). Lightly rub the peppers with oil and roast in the oven for about 30–40 minutes, turning once or twice, until slightly charred. Remove from the oven, cover with foil and allow to cool. When cool enough to handle, peel and deseed the peppers. Slice lengthwise into thin strips and set aside.

When the dough has risen, punch it down and knead again for 1–2 minutes. Divide the dough into two balls and, on a lightly-floured surface, roll each dough ball into a 10–12-in (25–30-cm) circle. Transfer to a baking tray and increase the oven temperature to 425°F (220°C). Spread each base with 2 Tbsp of the Salsa verde, sprinkle over the roasted peppers, olives and taleggio cheese. Season with black pepper, then cook in the middle of the oven for 10–12 minutes or until crispy and golden.

Serve immediately cut into wedges and sprinkled with basil leaves.

Vegetable tofu stir-fry

Colorful

Don't be restricted by the vegetables listed below. The great thing about a stir-fry is that you can use whatever you have in the fridge.

Serves 4

1 Tbsp **vegetable oil**
1 tsp **sesame oil (or extra tsp of vegetable oil)**
9oz (250g) **firm tofu**, cut into ½-in (1-cm) cubes
½ cup (150g) **baby corn**
½ cup (150g) **snow peas or fine green beans**
½ cup (150g) **tenderstem broccoli heads or bok choy**
1 **red pepper, cored,** deseeded and finely sliced
1 **carrot,** peeled and cut into julienne
1 Tbsp **soy sauce**
1½ Tbsp **oyster sauce**
Juice of 1 **lemon**
1 tsp finely grated peeled **fresh root ginger**
3 Tbsp **sweet chili sauce**
½ cup 75g **bean sprouts**
Small handful **fresh cilantro leaves,** chopped

In a large nonstick frying pan or wok, heat the vegetable oil and sesame oil until hot. Add the tofu and fry, turning frequently, until lightly golden. Remove using a slotted spoon and drain on paper towel.

To the remaining oil add all of the vegetables except the bean sprouts and stir-fry, stirring continuously, for 3–4 minutes or until vibrant in color. Add the soy sauce, oyster sauce, lemon juice, ginger and sweet chili sauce and mix to combine. Cook for a further 2–3 minutes or until the vegetables are almost cooked but still have a slight bite. Return the tofu to the pan with the bean sprouts and cilantro and mix to combine.

Serve immediately with extra cilantro sprinkled on top if liked.

Eggplant and okra curry

Spiced

Okra, a South American tapered green vegetable, is a natural thickener in soups and stews. Make sure the eggplant is really soft and almost mushy before serving.

Serves 4

2 large eggplants
1 Tbsp **vegetable oil**
1 **quantity Thai chili paste**
   (see page 166)
½ cup (100g) **okra,** cut into thirds
4 **large tomatoes, cut into large**
   chunks or wedges
1¾ cups (440 ml) **canned coconut**
   milk
Juice of 1 **lime**
1 large handful **fresh cilantro leaves**
Salt to taste
Freshly-cooked rice to serve
Extra chopped **cilantro, to garnish**

Preheat the oven to 400°F (200°C). Lightly coat the eggplants in a little oil and place on a baking tray. Bake for 20–25 minutes or until soft but still holding their shape.

Heat the oil in a large frying pan. Add the Thai chili paste and cook for 1–2 minutes until aromatic. Add the okra and cook for about 3 minutes. Then add the tomatoes and stir until they soften. Cut the eggplants into large chunks and add together with the coconut milk, reduce the heat and leave to simmer for 15–20 minutes or until the vegetables are tender. Remove from the heat and stir in the lime juice and cilantro and season with salt to taste. Serve with freshly cooked rice and chopped cilantro.

Eggplant and okra curry

Spiced sesame lentil patties with yogurt sauce
Aromatic

It may look like a lot of ingredients, but these patties really are quite simple to make. Be organized and have everything sliced, diced and weighed before you start cooking.

Makes 12

¾ cup (175g) **green lentils,** rinsed
2¼ cups (500ml) **cold water**
2 **bay leaves**
3 Tbsp **extra-virgin olive oil**
1 **large onion,** peeled and diced
2 **celery sticks,** diced
1-in (2.5-cm) **piece fresh root ginger,** peeled and grated
¼–½ tsp **crushed red pepper**
½ cup (100g) **sesame seeds**
1¼ cups (75g) **dried breadcrumbs**
1 tsp **ground cumin**
½ tsp **turmeric**
1 small bunch **flat-leaf parsley,** chopped
Juice of 1 **lemon**
1 tsp **salt, or to taste**

For the yogurt sauce:
Scant 1 cup (200g) **plain unsweetened yogurt**
1 tsp **ground cumin**
1 tsp **ground coriander**
Juice and zest of 1 **lemon**

Put the lentils in a large pan with the water and bay leaves. Bring to a boil, reduce the heat and simmer for 30–40 minutes or until cooked but not too soft. Remove from the heat, remove the bay leaves and drain. Set aside.

In a large heavy-based saucepan heat 2 Tbsp of the oil. Add the onion and celery and sauté for 2–3 minutes or until translucent but not browned. Stir in the ginger, crushed red pepper, sesame seeds, breadcrumbs, cumin, turmeric and parsley. Stir the breadcrumb mixture into the lentils. Blitz it briefly using a hand blender, or alternatively mash it together with a fork until the mixtures are amalgamated. Stir in the lemon juice and add salt to taste. Shape the mixture into 12 equal-sized patties.

Prepare the yogurt by combining it with the cumin, coriander and lemon juice and zest. Mix well then set aside.

Heat the remaining oil in a large nonstick frying pan and cook the patties for 2–3 minutes on each side or until golden and hot through. Serve the patties hot with the yogurt sauce on the side.

Wild mushroom and lentil ragù

Autumnal

A tiny French inn in late October; the proprietor has a basket brimming with wild mushrooms. That night we eat creamy lentils stirred with rich, earthy chanterelles. Sometimes what you read about in books does come true.

Serves 4–6

For the sauce:
1/4 cup (10g) **dried porcini mushrooms**
Scant 1 cup (200ml) **boiling water**
1 Tbsp **butter**
1 **small onion,** peeled and finely chopped
2 **garlic cloves,** peeled and finely chopped
Generous 1½ cups (150g) **mixed wild mushrooms,** roughly chopped
½ cup (100ml) **white wine**
⅔ cup (150ml) **heavy cream**
3 Tbsp grated **Parmesan cheese**
½ tsp **flaky sea salt,** or to taste
¼ tsp **cracked black pepper**

For the lentils:
Scant 1 cup (200g) **puy lentils**
1 **carrot,** peeled and cut in half
2 **garlic cloves,** peeled
1 **bay leaf**
1 **sprig fresh thyme**

To serve:
4–6 **slices crusty bread,** toasted
1 tsp **white truffle oil (optional)**
½ cup (10g) **flat-leaf parsley,** roughly chopped

Place the dried porcini mushrooms in a small bowl and cover with the boiling water. Leave to soak for 20 minutes. Drain through paper towel or a muslin-lined sieve, pushing the mushrooms with the back of a spoon. Reserve the liquid. Roughly chop the mushrooms and set aside.

Combine all of the ingredients for the lentils in a saucepan and cover with water. Bring to a boil, then reduce heat and simmer for 20–30 minutes, until lentils are tender. (Add more water if level drops below lentils.) Drain and discard carrot, garlic, thyme and bay leaf. Set lentils aside.

For the sauce, heat the butter in a frying pan over medium heat. Add the onion and garlic and stir occasionally for 5 minutes, until onions are soft. Add the porcini and wild mushrooms and stir occasionally for 5 minutes. Turn up the heat and add wine. When the bubbles subside, reduce the heat to medium-low and add ½ cup (100ml) of the porcini soaking liquid and cream. Simmer for 10 minutes until the sauce begins to thicken. Stir in the lentils, Parmesan cheese, salt and pepper.

Serve on toast, topped with a splash of truffle oil and a sprinkling of parsley.

Black bean and vegetable chili with cilantro-lime crème fraîche

Warming

This is "big bowl" food, intended to be eaten curled up in front of the TV with a lager and a large plate of tortillas.

Serves 4–6

For the chili:
4 Tbsp **olive oil**
2 **onions,** peeled and roughly chopped
2 **red peppers,** cored, deseeded and roughly chopped
3 **garlic cloves,** peeled and finely chopped
2 **chipotle chilies in adobo sauce,** roughly chopped
1 Tbsp **ground cumin**
1 Tbsp **ground coriander**
2 tsp **chili powder**
2 x 15-oz (425-g) **cans black beans,** drained and rinsed
2 x 15-oz (425-g) **cans crushed tomatoes**
¼ cup (75ml) **honey**
¼ cup (75ml) **cider vinegar**
1¾ cups (300g) **frozen sweetcorn**
1–2 cups (250–500ml) **vegetable stock,** depending on desired thickness

For the crème fraîche:
Scant 1 cup (200ml) **crème fraîche or sour cream**
3 Tbsp finely chopped **fresh cilantro**
Zest and juice of 1 **lime**

Heat the olive oil in a large soup pot over medium-high heat. Add the onions, peppers, garlic and chilies and cook, stirring occasionally, until beginning to soften, about 5 minutes. Add the cumin, ground coriander and chili powder and stir for 1 minute more. Add the beans, tomatoes, honey, cider vinegar, sweetcorn and enough stock just to cover. Simmer over a low heat for 15 minutes, stirring occasionally.

Combine the crème fraîche, fresh cilantro, and the lime juice and zest in a small bowl.

To serve, divide the chili between 4–6 warmed bowls, and top with a dollop of the crème fraîche.

Black bean and vegetable chili with cilantro-lime crème fraîche

Chickpea and cilantro cakes

Feisty

These feisty little cakes can be served as a starter with a dollop of crème fraîche and fresh cilantro or as a main course with a green salad and Rhubarb chutney (see page 163).

Makes 8 as a main course, 12 as a starter

Scant 1 cup (200g) **canned chickpeas,** drained and rinsed
1½ cups (150g) **freshly-cooked brown rice**
½ cup (50g) **oatmeal,** uncooked
1 tsp **coriander seeds**
1 tsp **cumin seeds**
3 Tbsp **peanut or vegetable oil**
1 **small onion,** peeled and finely chopped
1 **celery stalk,** finely chopped
2 **garlic cloves,** peeled and minced
1 **small red chili,** deseeded and finely chopped
1½ cups (20g) **fresh cilantro,** tops and roots separated, roots chopped
1 tsp **turmeric**
1 Tbsp **sea salt**
Zest and juice of 1 **lemon**
3 Tbsp **Greek yogurt, crème fraîche or sour cream**

Combine chickpeas, rice and oatmeal in a large bowl and set aside. Toast the coriander and cumin seeds in a dry frying pan over medium-high heat until the seeds are fragrant and begin to pop. Transfer to a pestle and mortar and grind until smooth.

Heat 1 Tbsp of oil in a frying pan over medium heat. Add the onion, celery, garlic, chili and chopped cilantro roots and sauté until soft, about 10 minutes. Stir in ground coriander, cumin seed and turmeric. Transfer the mixture to the bowl with the chickpeas, rice and oatmeal. Stir in the salt, lemon zest and juice, Greek yogurt and cilantro leaves. Place the mixture in a food processor and pulse until almost smooth.

Form the mixture into cakes – make either 8 or 12, depending on whether you are serving them as a starter or main course. Heat 2 Tbsp oil in a nonstick frying pan and heat until very hot. Fry the cakes for 3–4 minutes each side, or until browned. Serve immediately.

■ *The pre-cooked cakes can be frozen.*

Chickpea and cilantro cakes

Gado gado

Gado gado

Pure

Gado gado is an Indonesian vegetable dish topped with Sambal kacang (see page 164), a spicy peanut sauce. It's an incredibly versatile dish – serve the sauce with whatever vegetables you have at hand. The key is to employ a combination of textures and flavors, from steamed broccoli to fresh greens. To make the dish even more hearty, serve with potatoes or brown rice.

Serves 6

1 x 9-oz (250-g) **block firm tofu**
2 Tbsp **dark soy sauce**
Vegetable, seed or nut oil for frying
1 **large red onion,** peeled and sliced
4 **carrots,** peeled and sliced
1 **head cauliflower,** cut into florets
1 **head broccoli,** cut into florets
3 cups (200g) **snow peas**
1 **quantity Sambal kacang**
   (see page 164)

To garnish (optional):
Pea shoots
2 **hard-boiled eggs,** quartered

Place the tofu in a shallow bowl and put a small plate directly on top of the tofu. Put something heavy, like a can of soup, on the plate and leave for 20 minutes. (The weight of the can will force excess water from the tofu. Less water makes the tofu crispier when fried.) Drain away the excess water and chop the tofu into cubes. Return to the bowl and toss with the soy sauce. Leave to marinate for 20 minutes, tossing occasionally.

Heat 1 Tbsp oil in a frying pan over medium-high heat. Add the sliced onions and sauté, stirring occasionally, for 5 minutes. Reduce heat to low and leave to cook, stirring occasionally, for 30 minutes.

Heat 1 Tbsp oil in another frying pan over medium-high heat. Fry the marinated tofu for 2–3 minutes each side, until golden. Set aside and keep warm.

Bring a large pan of salted water to a gentle boil. Add the carrots and cook for 2 minutes. Add the cauliflower and cook for 2 minutes more. Add the broccoli and cook for a further 2 minutes. Add the snow peas and simmer for a further minute, then drain all the vegetables into a colander.

To serve, heat the Sambal kacang until hot (add water if too thick). Divide the vegetables and tofu among warmed plates and spoon the sauce on top. Garnish with pea shoots and hard-boiled eggs, if desired.

Poached egg with sautéed spinach and Jerusalem artichoke rösti

Warming

The Jerusalem artichokes give an interesting, nutty flavor to these potato cakes. Add a golden egg and wilted spinach for comfort food at its best.

Serves 4

12 medium (450g) **Jerusalem artichokes,** halved
Juice of 1 **lemon**
1 tsp **salt**
3 medium (450g) **red potatoes,** halved
Freshly ground black pepper to taste
1 Tbsp **all-purpose flour**
2 Tbsp **butter**
1 Tbsp **olive oil**
4 **large organic eggs**
1 tsp **white vinegar**
Spinach, rinsed, **to serve**
Sea salt and cracked pepper to taste

Preheat the oven to 425°F (220°C). Place the artichokes in a medium-sized saucepan with the lemon juice and ½ tsp salt. Bring to the boil and simmer for 6 minutes, until almost tender. Place the potatoes in a separate saucepan. Cover with water, add ½ tsp salt and bring to the boil. Simmer for 10 minutes until barely tender. Drain and cool. Grate the artichokes and potatoes into a bowl. Discard any loose potato skin. Add

pepper and combine the mixture with your fingers.

Shape the mixture into eight rounds 3¼in (8cm) wide and ½ in (1cm) thick. Press together and dust each lightly with flour.

Heat the butter and oil in a large frying pan until hot. Add the rounds and fry for 3–4 minutes each side, until golden. Set aside and keep warm.

Place the spinach in a frying pan over medium heat and stir until just wilted. Season to taste and set aside while you poach the eggs.

Pour 1½ in (4cm) of cold water into a large saucepan. Add the white vinegar and bring to a gentle simmer. Break eggs, one at a time, into a small bowl and slide into the water. Poach until the whites are firm but the yolks are still runny, about 2–3 minutes. Remove with a slotted spoon and place on paper towel to drain.

Arrange the rösti on four plates. Top with spinach and a poached egg. Season with sea salt and cracked pepper and serve immediately.

Poached egg with sautéed spinach and Jerusalem artichoke rösti

Vegetable Kao Soi with skinny fries

Vegetable Kao Soi with skinny fries

Wow

Kao Soi – also known as Thai Chiang Mai Curry noodles – is a classic dish from Northern Thailand featuring flavors from India and Thailand. We've replaced noodles with lots of colorful vegetables, and topped the dish with skinny sweet potato fries. A kao soi of sorts, shall we say?

Serves 4–6

2 Tbsp **vegetable or peanut oil**
2–3 Tbsp **Thai red curry paste**
2 tsp **curry powder**
1 tsp **turmeric**
2 x 14-oz (400-ml) cans **coconut milk**
1 cup (250ml) **vegetable stock**
1 **whole star anise**
1 **cinnamon stick**
1 tsp **brown sugar**
½ medium (500g) **cauliflower, cut into** florets
3 medium (500g) **zucchini, green or yellow,** chopped into ¾-in (2-cm) dice
1 cup (100g) **baby corn**
Juice of 1 **lime**
2 Tbsp **fish sauce, or to taste**

To garnish:
4 **green onions,** sliced diagonally
1 **sweet potato,** julienned
2¼ cups (500ml) **oil for frying**
2 **limes,** quartered

Heat the oil in a large wok or soup pot over medium-high heat. Add the curry paste, curry powder and turmeric and stir quickly until very fragrant, 1–2 minutes. Add the coconut milk, 1 cup (250ml) at a time, bringing it to a boil after each addition. Add the stock, star anise, cinnamon stick, brown sugar, cauliflower, zucchini and baby corn and simmer gently for 10 minutes or until vegetables are tender. Add lime juice and fish sauce to taste.

Heat the oil to very hot in a wok or large pan. Fry the julienned sweet potato in batches and drain on paper towel.

Spoon the curry into bowls. Garnish with green onions and lime quarters and crown with a tangle of skinny sweet potato fries.

Moroccan vegetable tagine

Hearty

A tagine is a slow-cooked dish or stew eaten throughout northern Africa and even southern Spain. They come in many different forms, often including meat or couscous as the main ingredient. I like to serve this particular tagine with a dollop of plain unsweetened yogurt on the side and a plateful of couscous if you're feeling extra hungry.

Serves 4–6

1 cup (150g) **dried apricots**
¼ cup (150ml) **boiling water**
2 Tbsp **olive oil**
1 **onion,** peeled and finely sliced
1½–2 Tbsp **harissa paste (depending on taste)**
½ tsp **ground cinnamon**
½ tsp **ground cumin**
2 **parsnips** (about 5oz/150g)**,** peeled and cut into ¾–1¼-in (2–3-cm) chunks
2 **carrots** (about 5oz/150g)**,** peeled and cut into ¾–1¼-in (2–3-cm) chunks
1 **large sweet potato** (about 10–12oz/300–350g)**,** peeled and cut into ¾–1¼-in (2–3-cm) chunks
1¼lb (600g) **pumpkin or butternut squash,** peeled and cut into 3-cm (1¼-in) chunks
14oz (400 g) **can chickpeas,** drained and rinsed
Salt to taste
½ cup (50g) **slivered almonds,** toasted
Small handful **fresh cilantro,** roughly chopped

Put the apricots in a small bowl and pour over the water. Leave to soak while you prepare the rest of the tagine.

In a large, heavy-based saucepan, heat the oil over medium heat. Add the onion and sauté for 2–3 minutes or until soft and translucent but not browned. Stir in the harissa, cinnamon and cumin. Then add the prepared vegetables and stir well until evenly coated. Cover and leave to cook over low-medium heat for 10 minutes, stirring occasionally.

Remove the lid and stir in the chickpeas, apricots and their soaking water and stir to combine. Cover again and continue to cook, stirring occasionally, for a further 20 minutes or until the vegetables are tender. Remove from the heat, season to taste and sprinkle over the almonds and cilantro. Stir gently to combine then serve immediately.

Moroccan vegetable tagine

Mushroom and tarragon Wellington

Earthy

This is a great vegetarian alternative for Christmas Day, with its almost meat-like consistency and substantial qualities.

Serves 4–6

2 Tbsp **olive oil**
1 **large onion,** peeled and chopped
2 **garlic cloves,** peeled and crushed
1¼ cups (30g) **porcini mushrooms,** soaked in warm (boiled) water for at least 20 minutes
3 cups (300g) **chestnut mushrooms,** sliced
1 handful **fresh tarragon,** chopped
1 Tbsp **soy sauce**
1 Tbsp **Marsala**
Salt and freshly ground black pepper
1 cup (150g) **unsalted cashew nuts**
¾ cup (100g) **blanched almonds**
1⅔ cups (100g) **whole-wheat breadcrumbs**
13 oz (375g) **puff pastry**
1 **egg to glaze**

In a large frying pan heat 1 Tbsp of oil. Sauté the onion for 2–3 minutes or until soft and translucent but not browned. Add 1 clove of garlic and sauté for a further minute. Set aside.

Drain the porcini, reserving the liquid, and roughly chop. Heat the remaining oil in a large frying pan and add the chestnut mushrooms. Sauté until soft, add the chopped porcini and remaining garlic clove and continue cooking until lightly golden. Stir in the tarragon together with the soy sauce and Marsala and season generously with salt and pepper. Continue to cook until the liquid has evaporated.

Put the cashew nuts and almonds in a food processor, together with 1 Tbsp of the porcini soaking liquid. Process until ground, then add the onion and mushroom mixtures and process again. Add the breadcrumbs and process briefly, adding extra porcini soaking liquid if necessary. Set aside while you prepare the pastry.

Preheat the oven to 400°F (200°C). On a lightly-floured surface, roll the pastry into a rectangle about 11¾ x 8½in (30 x 22cm). Place on a lightly-oiled baking tray. Place the mushroom mixture in an even log shape down the center of the pastry. At ¾-in (2-cm) intervals, cut the pastry at 45-degree angles on both sides, from the mushroom mixture out to the edge of the pastry. Alternately lift the cut strips so that they lie across the mushroom mixture and overlap to form a plait.

Brush the pastry all over with the beaten egg and cook in the oven for 40–50 minutes or until golden brown and hot through. Serve cut into slices.

Perogies, Canadian-Ukrainian style

Comforting

I could eat these crescent-shaped dumplings from the Ukraine every day.
Greg Lypowy, my Ukrainian-Canadian neighbor, shared this recipe with me.

Makes 45–50

For the dough:
4 cups (450g) **all-purpose flour**
1½ tsp **salt**
1 **egg**
Scant 1 cup (200ml) **cold water**
4 tsp oil or 2 Tbsp **sour cream**

For the filling:
2 Tbsp **butter**
½ **large onion**, finely chopped
3 **garlic cloves**, peeled and minced
5 medium (750g) **potatoes**, mashed
7oz (200g) **Cheddar cheese**, grated
1 tsp **salt**
½ tsp **pepper**

For the topping:
3 Tbsp **butter**
1 **large onion**, peeled and chopped
½ tsp **salt**
Scant 1 cup (200ml) **sour cream**
4 Tbsp roughly chopped **fresh dill**

To make the dough, combine the flour and salt in a large bowl. In a separate bowl, beat together the egg, water and oil. Stir the liquid into the flour mixture until you have a soft dough that holds together in a ball. If the dough is too dry, add a little more water. Allow to rest for 30 minutes.

Heat the butter in a large frying pan. Add the onion and fry for 10 minutes, until golden. Stir in the garlic and cook for a few minutes more. Set aside.

Combine the mashed potatoes, cheese, onion and garlic, salt and pepper in a bowl and then set aside.

On a lightly floured surface roll half the dough out until it's ⅛in (3mm) thick. Cut circles using a 3-in (7.5-cm) cookie cutter. Place a heaped teaspoon of the filling in the center of each circle. Dip your finger into water and then run it around the edge of circle. Fold in half, pinching to seal. Place the perogies on a lined baking tray. Continue until you have used all of the dough.

To prepare the topping, heat the butter in a large frying pan. Add the onion and fry for 10 minutes until golden. Add the salt and keep warm.

Bring a large pan of water to a boil. Drop the perogies into the water. Remove with a slotted spoon when they rise to the surface. Transfer directly to the pan with the onions. Give the perogies and onions a gentle stir, then serve together with a spoonful of sour cream and a sprinkling of dill.

Mushroom, wild rice and ale pie

Mushroom, wild rice and ale pie

Comforting

This is a real Sunday lunch treat for anyone, vegetarian or not. The rice adds a unique texture to complement the mushrooms.

Serves 4–6

4 small (200g) **Portobello mushrooms**
2½ cups (200g) **chestnut mushrooms**
2½ cups (200g) **wild mushrooms (e.g. shiitake, enoki, chanterelle)**
½ stick (30g) **butter**
2 Tbsp **olive oil**
1 **onion,** peeled and finely chopped
1½ cups (30g) **dried porcini,** soaked in scant 1 cup (200ml) boiling water
3 **garlic cloves,** peeled and crushed
1 medium (250g) **celeriac,** peeled and cut into ½-in (1-cm) cubes
2 Tbsp **all-purpose flour**
⅔ cup (150ml) **dark ale**
1¼ cups (100g) **freshly-cooked wild rice**
Salt and freshly ground black pepper
1lb 2oz (500g) **puff pastry**
2 Tbsp chopped **fresh thyme leaves or flat-leaf parsley**
Milk to glaze

Rinse the mushrooms and wipe clean. Cut each Portobello mushroom into eight, the chestnut mushrooms in half and the rest as appropriate.

Heat the butter and oil in a large heavy-based saucepan and sauté the onion until soft and translucent but not brown. Drain the porcini, reserving the liquid and chop finely. Add the garlic and chopped porcini to the pan and sauté for 1 minute before adding the celeriac. Stir to coat in the oil then add the mushrooms. Keep the heat high and cook the mushrooms, stirring frequently until softened.

Gradually add the flour and stir gently until evenly combined. Add the ale and reserved porcini liquid and stir, over a high heat, until the sauce thickens before stirring in the rice. Season to taste. Continue to cook for a further 2–3 minutes then remove from the heat and cool to room temperature.

Preheat the oven to 425°F (220°C). Roll out two-thirds of the pastry on a lightly-floured surface to ⅛in (3–4mm) thick and use to line a 9-in (23-cm) rimmed pie pan. Stir the thyme or parsley into the mushroom mixture then spoon into the pie pan. Brush the rim of the pastry with milk.

Roll out the remaining pastry and lay it over the top of the mushroom mixture and trim to fit. Using your fingers or a fork, pinch the pie edge together. Make a cross in the center of the pie and brush with a little more milk. Cook in the oven for 40–50 minutes or until golden and crispy. Serve immediately.

Parmesan polenta with sautéed mixed mushrooms and wild garlic

Parmesan polenta with sautéed mixed mushrooms and wild garlic

Sophisticated

Although this dish is very rustic in theory, it will impress any guest with its sophisticated appearance. It's all about choosing an interesting selection of mushrooms to give it the wow factor.

Serves 4

4½ cups (1L) **vegetable stock or cold water**
1 tsp **salt**
Freshly ground black pepper
1½ cups (250g) **polenta or coarse maize**
2oz (50g) **Parmesan cheese,** grated
1 Tbsp (15g) **butter**
2 Tbsp **extra-virgin olive oil**
6¼ cups (500g) **mixed mushrooms (e.g. shiitake, enoki, chestnut)**
½ cup (50g) **wild garlic shoots**
1 Tbsp chopped **fresh thyme leaves**
2 Tbsp **marsala**
Freshly ground black pepper

Bring the stock to a boil in a large saucepan. Season with salt and pepper. Add the polenta in a thin stream, whisking it continuously into the boiling liquid. As soon as it comes to a boil, reduce the heat and swap your whisk for a wooden spoon. Stir in the Parmesan and simmer the polenta, stirring regularly, for about 10 minutes or until thick and porridge-like. Be careful as it will splatter volcanically.

Line a baking tray with parchment paper and quickly spread the polenta evenly over it to about ½in (1.5cm) thick. Use a dampened pallet knife to spread it evenly and smooth the surface. Leave the polenta to cool and set for at least 30 minutes. This can be prepared up to two days in advance to this stage.

Cut the polenta into eight triangles. Heat 1 Tbsp of the oil in a large non-stick frying pan and fry the polenta on both sides until lightly golden and crispy. Remove from the pan and place two triangles on each plate.

Put the remaining oil and butter in the same pan and heat to hot. Add the mushrooms and sauté, stirring briskly until golden and cooked. Stir in the wild garlic shoots, then add the marsala. Cook until the marsala has evaporated, and season to taste. Divide the mushrooms evenly among the plates, placing them on top of the polenta, and serve immediately.

Side dishes

One of the hardest parts of writing this book was relegating recipes into the "sides" chapter. It makes us feel bad. We go on about how special vegetables are; how they are so rarely featured in the spotlight; how they are always the bridesmaid, never the bride. And here we go and put them … on the side.

We're tempted to call the chapter "accessories." These recipes are the part of the ensemble that aren't absolutely necessary, but they do pull everything together in a unique, fabulous way. Yes, that's better.

This chapter is intended to serve as a resource for seasonal cooking. Flip to these pages when you come home from the market with a basket full of artichokes, peppers, cabbage, Brussels sprouts, celeriac or even bright green tomatoes. They weren't on your list, but you had to have them. We understand. It's the case with all great accessories.

Egg fried rice

Leftovers

This is a great way to use up leftover vegetables. If you're not after a strictly vegetarian version, then try adding a few shrimp or finely sliced chicken pieces as well.

Serves 4–6

2 Tbsp **vegetable oil**
1 **celery stalk,** finely sliced
1 **onion,** peeled and finely chopped
4–5 leaves **Savoy cabbage,** finely shredded
1 **red pepper,** cored, deseeded and finely diced
2 **eggs**
1 Tbsp **soy sauce**
1¼ cups (150g) **white rice,** cooked as per package instructions
3 **green onions, finely sliced**
Salt to taste

Heat the oil in a large nonstick frying pan or wok. Add the celery and onion and gently sauté until soft. Add the cabbage and red pepper and sauté for a further couple of minutes.

Gently beat the eggs with the soy sauce and add to the frying pan. Keep stirring over a high heat until the egg is set. Add the rice and half of the spring onions and continue cooking until hot through, about 3–4 minutes. Remove from the heat and taste, adding more salt if needed. Stir in the remaining green onions and serve immediately.

Sautéed mixed greens

Piquant

You can give this dish more of an Asian or Mediterranean twist depending on the vegetables you choose. Use sesame oil for an Asian accompaniment.

Serves 4–6

1½lb (700g) **mixed greens (e.g. tenderstem or purple sprouting broccoli, spinach, endive, Swiss chard, bok choy)**
2 Tbsp **extra-virgin olive oil, plus extra for serving**
2 **garlic cloves, peeled and crushed**
½ tsp **crushed red pepper**
Sea salt and freshly ground black pepper
1 Tbsp **sesame seeds, toasted**

Trim the greens and wash them thoroughly. Bring a large pan of salted water to a boil and simmer the greens until just tender and bright green in color. Drain and rinse briefly under cold running water and squeeze out as much excess water as possible.

In a large nonstick frying pan or wok heat the oil. Add the garlic, crushed red pepper and greens and sauté for 2–3 minutes or until heated through. Season to taste and then toss through the sesame seeds and serve immediately.

Roasted red beets with tarragon dressing

Vibrant

Ruby-red beets add sparkle to any meal. Serve alongside a main dish or all by itself upon a bed of couscous or quinoa and a handful of fresh greens. You can't go wrong either way.

Serves 4

12oz (350g) **red beets,** tops removed

For the dressing:
4 Tbsp **extra-virgin olive oil**
2 Tbsp **red wine vinegar**
3 rounded Tbsp chopped **fresh tarragon**
½ tsp **Dijon mustard**
1 tsp **superfine sugar**
1 **garlic clove,** peeled and minced
Sea salt and cracked black pepper to taste

Preheat the oven to 400°F (200°C). Wash the red beets and pat dry. Place them in foil and wrap up the ends, scrunching to seal. Place the package in the oven and roast for 1 hour, or until the beets are tender when pierced with a knife. Open the package and set aside to cool.

When the beets are cool enough to handle, rub off the loose skins and slice each beet into quarters or in half, depending on its size. Arrange the beets on a serving platter.

Combine the dressing ingredients (reserve 1 Tbsp tarragon for garnish) in a small bowl and whisk well to combine. Pour over the warm beets and toss gently to coat. Garnish with remaining tarragon and season with sea salt and cracked black pepper to taste.

Roasted red beets with tarragon dressing

Sautéed Savoy cabbage

Sweet

Cabbage is such a lovely, simple winter accompaniment to any meal. Substitute with other local varieties if you prefer.

Serves 6

1 **head Savoy cabbage**
2 Tbsp **olive oil**
2 Tbsp **butter**
1 **small red onion**, peeled and finely sliced
½ tsp **sea salt**
Cracked black pepper to taste

Separate the cabbage leaves then cut out the central core and discard. Finely slice remaining leaves.

Heat the oil and butter in a saucepan over medium heat. Add the onion and salt, then reduce the heat to low. Cook, stirring occasionally, until the onions begin to soften, about 20 minutes. (This step can be done several hours before serving. Leave onions in the pan then reheat before continuing.)

Add the sliced cabbage, stir well, and cook for a further 25 minutes, until the cabbage is very soft. Add a little water during cooking, if necessary, to keep the cabbage moist.

Serve with cracked black pepper to taste.

Sautéed spinach with garlic, figs and pine nuts

Persian

This quick-to-prepare combination of flavors is delicious as a side dish, all alone or as a topping for the Jerusalem artichoke rösti (see page 56).

Serves 4–6

1 Tbsp **olive oil**
2 **shallots,** peeled and finely chopped
2 **garlic cloves,** peeled and finely chopped
8 cups (1kg) **spinach,** roughly chopped
3 Tbsp **pine nuts,** toasted
4 **fresh figs,** roughly chopped
1 Tbsp **balsamic vinegar**
Sea salt and cracked black pepper to taste

Heat the oil in a large frying pan over high heat. Add the shallots and garlic and quickly stir-fry for 1 minute, until fragrant. Add the spinach. Stir constantly for 1–2 minutes, until wilted.

Remove from the heat and toss in the toasted pine nuts and figs. Transfer to a serving dish and drizzle with vinegar. Season to taste. Serve immediately.

■ *Good balsamic vinegar is thick, almost syrupy, and expensive – but a little goes a long way. If the regular, inexpensive variety is on hand, there is a way to posh it up a little: heat ¼ cup (75ml) in a small saucepan and simmer until reduced by half.*

Ruby coleslaw

Crunchy

There is nothing posh about this side dish – it's just simple, crunchy, satisfying fare. Serve it with sandwiches, grilled vegetables, baked beans or all by itself.

Makes 2¼ cups (500ml)

½ **head cabbage,** approx 18oz (500g)
2–3 **carrots,** peeled
2 Tbsp **apple cider vinegar**
4 Tbsp **mayonnaise**
4 Tbsp **superfine sugar**
1 tsp **salt**
1 tsp **cracked black pepper**

To serve:
2 Tbsp **mixed seeds (try pumpkin, sunflower, hemp, linseed or fennel)**
1/2 cup (50g) **radish sprouts**

Peel outer layer from cabbage and remove core. Shred both the cabbage and the carrot into fine slices. (This is easiest in a food processor using the grating blade. If that's not an option, use a knife or a sharp grater.) Transfer to a bowl and set aside.

Combine the remaining ingredients in a small bowl. Pour over the grated carrot and cabbage and toss to coat. Cover and refrigerate for at least 1 hour or up to 6 hours.

Before serving, top the coleslaw with mixed seeds and sprouts.

Ruby coleslaw

Caramelized shallots and Brussels sprouts

Caramelized shallots and Brussels sprouts

Sticky

This dish is a perfect accompaniment to Christmas dinner. If you're not a vegetarian, this dish is divine with just a little bit of bacon.

Serves 4–6

4½ cups (500g) **Brussels sprouts,** trimmed
1 Tbsp **olive oil**
1 Tbsp (15g) **butter**
1lb 2oz (500g) **shallots,** peeled
1 Tbsp **superfine sugar**
3 **bay leaves**
2 Tbsp **sherry vinegar or balsamic vinegar**
1 Tbsp chopped **fresh chives**

Bring a large pan of water to a boil and cook the Brussels sprouts until a sharp knife just goes through to the center when tested. Make sure that you do not overcook the sprouts.
Drain and set aside.

Heat the oil and butter in a large nonstick, ovenproof frying pan. Add the shallots and cook over a moderate heat, stirring frequently until well coated. Add the sugar and bay leaves and continue cooking and stirring until lightly golden. Increase the heat and add the vinegar, allowing it to bubble up and the majority to evaporate. Stir until the liquid thickens and is slightly syrupy. Add the sprouts and stir to combine.

Transfer the frying pan and its contents to the oven and cook for a further 10–15 minutes or until golden and caramelized. Serve immediately sprinkled with chopped chives.

Sesame miso spinach

Japanese

Stir some cooked tofu through the spinach for a more substantial side dish.

Serves 2

5 cups (225g) **spinach leaves**
1½ Tbsp **fermented miso paste (soybean paste)**
1 Tbsp **boiling water**
½ tsp **soy sauce**
½ tsp grated, peeled **fresh root ginger**
1 Tbsp **sesame seeds,** toasted

Wash the spinach and remove any tough stalks.

Combine the miso paste, water, soy sauce and ginger in a bowl to form a smooth, thick sauce.

Blanch the spinach then toss through the sauce. Sprinkle with sesame seeds and serve immediately.

Baby artichokes, Tuscan-style

Simple

I wrote this recipe down while sitting at Cousin Betsy's kitchen table in Tuscany. It was early spring and the local farmers' market was filled with baby artichokes. The artichokes can either be served as directed below or tossed through small ear-shaped pasta and topped with plenty of grated Parmesan, as Betsy does.

Serves 4

1 **lemon**
10 **baby artichokes**
4 Tbsp **olive oil**
3 **garlic cloves,** peeled and finely chopped
1 **sprig fresh rosemary,** needles finely chopped
¼ tsp **sea salt**
Grated Parmesan cheese

Fill a mixing bowl with cold water. Cut the lemon in half and squeeze the juice into the water. Put the squeezed lemon halves in the water as well.

Peel dry petals from the artichokes, then peel the outer layer of stalks, until all "stingy bits" are removed and the tender center is visible. Chop the top quarter off each bud and discard. Roughly chop the remaining heart and stalk. Place in the lemon water – this will keep the artichokes from discoloring.

Heat the olive oil in a saucepan over medium heat. Add the garlic and rosemary and stir for 1 minute. Drain the artichoke pieces and pat dry with paper towel. Add to the pan and stir to coat. Put the lid on, turn the heat to low and simmer until tender.

To serve, sprinkle with the salt and Parmesan cheese.

Braised fennel and garlic

Delicate

What might look like braised onion is in fact an intricate, delicate, anise-infused side dish that everyone will love.

Serves 4

2 **fennel bulbs,** tops trimmed
4 Tbsp **olive oil**
2 **cloves garlic,** peeled and squashed
½ cup (125ml) **white wine**
**Sea salt and freshly ground black
    pepper to taste**

Using a vegetable peeler, peel off any imperfections from fennel bulbs and then cut bulbs into quarters.

Heat oil in a large frying pan over medium-high heat. Choose a pan that has a tight-fitting lid, but leave the lid off for now. Add the garlic cloves and then the fennel quarters and cook, stirring occasionally, until golden brown, about 15 minutes.

Add wine, stir, then cover and simmer for 30 minutes over low heat. Add salt and pepper to taste. Serve sprinkled with extra fennel fronds.

Braised fennel and garlic

Fennel and celery bake

Creamy

This is great Sunday night fare in accompaniment to our Wild mushroom and lentil ragù recipe (see page 49). A good, hearty, warming meal.

Serves 4–6

1lb 2oz (500g) trimmed **fennel**
3 **celery stalks**
1 **bay leaf**
⅔ cup (150ml) **heavy cream**
⅔ cup (150ml) **milk**
1 tsp **Dijon mustard**
1 **garlic clove,** peeled and crushed
Salt and freshly ground black pepper
¾ cup (50g) **grated Parmesan or Gruyère cheese**
Small handful **fresh parsley,** chopped
1 cup (50g) **fresh whole-wheat breadcrumbs**
Extra-virgin olive oil

Preheat the oven to 400°F (200°C). Cut the fennel in half lengthwise then cut each half into 3–4 wedges depending on its size. Trim the ends off the celery and cut into 1-in (2.5-cm) pieces, on the diagonal.

Bring a pan of water to a boil. Add the fennel and bay leaf and simmer for 3 minutes, then add the celery and simmer for a further 2 minutes. Drain and set aside in a baking or gratin dish.

In a batter bowl, combine the cream, milk, mustard and garlic, season generously with the salt and black pepper and pour over the fennel and celery. In a separate bowl combine the Parmesan, parsley and breadcrumbs and sprinkle evenly over the mixture. Drizzle with olive oil.

Bake in the preheated oven for 25–30 minutes or until hot through and golden on top. Serve warm. Remove the bay leaf while serving.

Celeriac mash

Unique

Celeriac, also called celery root, is a variety of celery cultivated for its root, not its stalk. It has a unique flavor, somewhat like celery or parsley. It's ugly but tasty!

Serves 4–6

1 **celeriac** (about 2¼lb/1kg)**, peeled and chopped into 1¼-in (3-cm) dice
1 **fat garlic clove,** peeled
1 Tbsp **sea salt**
2 **small red potatoes,** peeled and chopped into 1¼-in (3-cm) dice
⅔ cup (150ml) **light cream**
4 Tbsp **unsalted butter**
Salt and freshly ground black pepper to taste

Place celeriac and garlic in a saucepan and add just enough cold water to cover. Bring to a boil and add the salt and potatoes. Simmer until the vegetables are soft and tender, about 10 minutes. Drain.

Add the cream and butter to the vegetables, then blend well with a hand blender (or transfer to a food processor) until smooth. Add salt and pepper to taste.

Red lentil dhal

Spiced

Dhal, the Indian word for "lentils," are a mainstay of Indian and Pakistani cuisine. You will find dhal prepared in varying ways depending on the region you are in, but all will be delicately spiced and slightly stew-like in texture.

Serves 4–6

Scant 1 cup (200g) **red lentils**
2 Tbsp **vegetable or sunflower oil**
1 **onion,** peeled and finely chopped
2 **garlic cloves,** peeled and finely chopped
1-in (2.5-cm) **piece fresh root ginger,** peeled and grated
½ tsp **crushed red pepper**
1 Tbsp **tomato paste**
2 **large tomatoes,** roughly chopped
2¼ cups (500ml) **boiling water**
Salt to taste
Juice of 1 **lemon**
1 large handful **fresh cilantro,** finely chopped

For the spice mix:
2 tsp **mustard seeds**
1 tsp **fenugreek**
2 tsp **coriander seeds**
2 tsp **cumin seeds**

2–3 **long red chilies,** roasted, to serve

Rinse the lentils and leave to soak in cold water while you prepare the spice mix.

In a small, dry pan heat all the spices until aromatic and starting to pop, about 1–2 minutes. Transfer to a mortar and pestle or spice grinder and grind to a fine powder. Set aside.

In a large nonstick saucepan heat the oil. Add the onion and sauté until translucent but not browned. Stir in the garlic, ginger and crushed red pepper. Then add the spice mix and sauté for a further 30 seconds. Stir in the tomato paste, chopped tomatoes and the water. Drain the lentils and add to the pan. Bring to a boil and simmer, stirring occasionally, for about 40–45 minutes or until the lentils are cooked and most of the liquid has been absorbed. Remove from the heat, season to taste with salt and stir in the lemon juice and cilantro. Serve warm, garnished with roasted red chilies.

■ *To roast whole red chilies, simply brush lightly with oil, place on a baking sheet and place in a medium oven (350°F) for 10-15 minutes. Allow to cool before handling.*

Red lentil dhal

Winter vegetable colcannon
Comforting

Colcannon is simply an Irish version of mashed potato and cabbage. We have added carrots for a hint of color and for additional nutrients but you could also use turnips or parsnips.

Serves 6

5 (700g) **potatoes,** peeled, boiled and mashed
½ head (250g) **Savoy cabbage or kale,** finely shredded and cooked
3 medium (250g) **carrots,** peeled, boiled and mashed
½ cup (100ml) **heavy cream**
½ cup (100ml) **milk**
½ stick (50g) **butter,** melted
Salt and freshly ground black pepper
2 Tbsp **chopped fresh chives**

In a large bowl or saucepan, combine the mashed potato, cooked cabbage and mashed carrots. Mix well to combine.

Heat the cream and milk in a heavy-based saucepan until it just comes to a boil then remove from the heat. Pour the liquid over the mash, add the butter and mash everything together. Season generously with salt and pepper to taste, transfer to a warm serving bowl and serve sprinkled with chopped chives.

Fried green tomatoes

Comforting

This is what Grammy Brownlee always does with any unripened green tomatoes. She serves it as a late-night snack, over toast. Delish!

Serves 2

1 Tbsp **butter**
2 **large, green tomatoes,** thickly
   sliced
¼ tsp **baking powder**
¼ tsp **salt**
¼ cup (50ml) **heavy cream**
Freshly ground black pepper
   to taste
Toast to serve

Heat the butter in a large frying pan over medium heat. Add the tomato slices, overlapping a little if necessary. Fry for 2–3 minutes per side, until golden brown. Sprinkle with baking powder and salt, and add the cream and stir. The tomatoes will become soft and mushy. Season with black pepper.

Serve over thick, buttered (in my grandmother's case, molasses brown bread) toast.

Cauliflower with toasted spices

Bold

This spicy, intriguing side dish is the perfect accompaniment to many Indian dishes. Try with the Red lentil dhal on page 86.

Serves 4

3 Tbsp **vegetable oil**
1 tsp **cumin seeds**
1 tsp **coriander seeds**
1 tsp **mustard seeds**
1 tsp **black onion seeds**
1 **dried hot pepper**
2 **shallots,** peeled and finely chopped
1 **clove garlic,** peeled and finely chopped
½ tsp **turmeric**
1 tsp **salt**
1 **cauliflower,** approximately 2lb (1kg), cut into florets

Heat oil in a large frying pan (choose one that has a tight-fitting lid, but leave the lid off for now) over medium-high heat. Add all the seeds and stir until fragrant, about 1 minute.

Add dried pepper and fry for a further minute. Stir in the shallots, garlic, turmeric and salt and fry for 5 minutes more.

Add the cauliflower and stir so that the florets are coated in the spicy seed mixture. Cover and lower heat. Cook, stirring occasionally, for 10–15 minutes or until cauliflower is tender. Serve immediately.

Cauliflower with toasted spices

Soups and salads

There is never a time when one should be without greens, regardless of locale. My friend Gillian used to sprout her own greens while away on weeklong canoe trips in northern Ontario. She would sprinkle seeds into a jar, soak them in water, rinse and rinse again. Within a few days, little sprouts would emerge, just in time to add nourishment to her week-old packed provisions.

Gillian is still a sprouter. She sprouts chickpeas, mung beans, clover and adzuki beans. She sprinkles them over salads, sandwiches, soups and dips. The variety of seeds available is endless. Who would have thought the pure essence of broccoli could be captured in a teeny, tiny leaf?

And then there are larger leaves. There is still a place for old faithfuls such as Romaine, but think of peppery arugula, tender baby spinach, soft butter lettuce, sharp watercress or pea shoots, if you're lucky enough to find them.

Explore the world of greens, even when you're in a canoe. But save the soup-making for dry land.

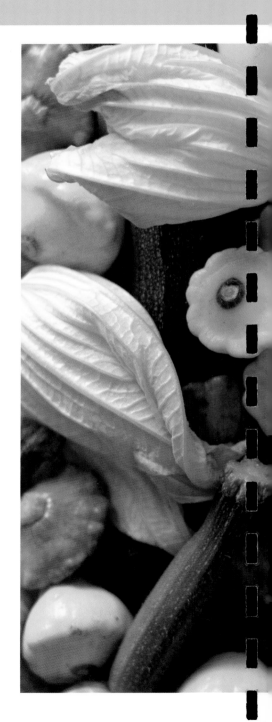

Butternut squash soup with blue cheese and walnuts

Intricate

Dark brown butter, blue cheese, fresh herbs and walnuts take a standard, silky soup to a more mature, complex level.

Serves 4

4 Tbsp **butter**
1 **small onion**, peeled and chopped
1 tsp **salt + extra to taste**
1 large (750g) **butternut squash,** peeled and chopped into chunks
2 **cloves**
¼ tsp **freshly grated nutmeg**
Freshly ground black pepper to taste
Generous 3 cups (750ml) **vegetable stock**
6 Tbsp toasted, chopped **walnuts**
6 Tbsp crumbled **blue cheese**
1 Tbsp chopped **fresh flat-leaf parsley**

Heat 1 Tbsp of the butter in a large soup pot over medium heat. Add the onion and cook, stirring occasionally, until softened. Add the tsp of salt, squash, cloves and nutmeg and cook the mixture, covered, over a low heat for 10–15 minutes or until the squash is tender. Add the stock and simmer for a further 15 minutes.

Purée the soup (transfer to a food processor in stages or use a hand blender), then strain the soup back into the pot. Add salt and pepper to taste.

Heat the remaining butter in a frying pan until the butter is nut-brown. Add the toasted nuts and stir to coat.

Ladle the soup into four bowls and garnish with the browned butter, walnuts, blue cheese and parsley.

Butternut squash soup with blue cheese and walnuts

Warm spinach, caramelized onion, feta and pecan salad

Warm spinach, caramelized onion, feta and pecan salad

Warming

Isn't it wonderful when a simple yet impressive salad can be made with such little effort? The onions can be caramelized a few hours in advance and left in the pan. Reheat before serving, add the pecans, spinach and feta, and there you have it. A warming, winter salad.

Serves 4

4 Tbsp **butter**
2 medium (about 10oz/300g) **onions,** peeled and sliced
½ tsp **anchovy paste (optional)**
½ cup (50g) **pecans,** roughly chopped
4oz (100g) **feta,** crumbled
5 cups (200g) **spinach,** washed and patted dry
Sea salt and freshly ground black pepper to taste

Melt the butter in a large frying pan over medium heat. Add the onions and stir to coat. (It seems like a lot of butter, but it constitutes the bulk of the dressing for the salad. Add more butter or a splash of olive oil if onions become dry.) Turn heat down to medium-low and leave onions to cook, stirring occasionally, for 30–40 minutes, until soft and golden.

Stir in the anchovy paste, if using, and the pecans and feta. Stir to coat. Now carefully fold the spinach into the onions. When the leaves are warm and coated, but not yet wilted, remove the pan from the heat.

Serve with a sprinkling of sea salt and freshly ground black pepper to taste.

Cauliflower soup with sourdough and pumpkin-seed croutons

Cauliflower soup with sourdough and pumpkin-seed croutons

Comforting

I don't think the humble cauliflower gets the high praise it deserves. This soup is so simple that its true colors can shine through. The croutons on top add texture and color to a dish that could otherwise look rather bland.

Serves 4

1 Tbsp (15g) **butter**
2 Tbsp **olive oil**
2 **leeks, sliced**
2 **bay leaves**
Salt and freshly ground pepper
1 **large head cauliflower,** cut into florets
5–5½ cups (1–1.25L) **hot vegetable stock**
2–3 **slices sourdough bread**
3 Tbsp **pumpkin seeds**
Pinch ground nutmeg
½ cup (100ml) **crème fraîche +**
extra to serve (optional)
Small handful fresh parsley,
chopped

Preheat the oven to 350°F (180°C). Heat the butter and 1 Tbsp of the oil in a large saucepan or stockpot. Add the leek and bay leaves and let it sweat over medium heat for about 5 minutes or until soft. Season generously with salt and freshly ground black pepper, then add the cauliflower and stock. Bring to a boil, reduce the heat and simmer for about 20 minutes or until the cauliflower is very soft.

While the soup is simmering prepare the croûtons. Remove the crusts from the sourdough bread and crumble into smallish pieces onto a baking tray. Add the pumpkin seeds, remaining Tbsp of oil and salt and pepper. Bake the croûtons for about 6–8 minutes, stirring once or twice to maintain even browning. Set aside until ready to serve.

When the soup is cooked, remove from the heat and blend to a smooth consistency. Return the soup to the heat and stir in the nutmeg, crème fraîche and half of the chopped parsley.

Serve bowls of soup topped with the croûton mix and parsley. Adding an extra dollop of crème fraîche always goes down well.

■ *Blue cheese is a delicious addition to this soup. Try crumbling it on top in addition to the croûtons, or stir it through with the cream so it melts in for a more even flavor.*

Edamame salad with minty vinaigrette

Tangy

Every mother raises her arms with joy when she discovers her children eating healthy food as if it were candy. That's what happened at my house with edamame, otherwise known by their less sexy name: soybeans. They are sold frozen in their pods or already shelled. This salad takes advantage of the latter – not as much fun for three-year-olds, but much easier in a salad.

Serves 4

1¼ cups (300g) **shelled edamame beans**
2 **organic lemons**
½ cup (10g) **fresh mint leaves,** finely sliced
3 Tbsp **olive oil**
2 Tbsp **rice vinegar**
2 Tbsp **honey**
Sea salt and black pepper to taste

Boil the edamame beans according to the directions on the package. Drain, rinse well under cold water, transfer to a serving bowl and cool to room temperature.

Zest both lemons, then juice only one. In a small bowl, combine the zest of 1 lemon, the juice of 1 lemon, half the mint leaves, oil, vinegar and honey. Add salt and pepper to taste.

Toss the dressing with the edamame beans, along with the remaining lemon zest and mint leaves. Serve immediately.

Fennel, Jerusalem artichoke and lentil salad

Wholesome

The Jerusalem artichoke is neither from Jerusalem nor is it actually an artichoke! Their mildly sweet and smoky flavor is delicious in salads – either cooked or raw, and very thinly sliced.

Serves 4–6

Scant 1 cup (200g) **green or puy lentils**
2½ cups (600ml) **cold water**
8 (350g) **Jerusalem artichokes,** cut into ½-in (1-cm) slices
2 **fennel bulbs,** quartered lengthwise and cut into small pieces
2 Tbsp **olive oil**

For the dressing
2 Tbsp **extra-virgin olive oil**
Juice of 1 **lemon**
2 tsp **Dijon mustard**
Small bunch **fresh flat-leaf parsley,** chopped
Salt and freshly ground pepper

Rinse the lentils then put them in a saucepan with the water and bring to a gentle simmer. Allow to simmer for about 30 minutes or until tender but not mushy.

Preheat the oven to 400°F (200°C). Put the Jerusalem artichokes and fennel in a roasting pan and cover with the oil. Roast in the hot oven for about 30 minutes, turning occasionally.

When the lentils are cooked, drain off any excess liquid. Combine the dressing ingredients in a small bowl and pour over the still-warm lentils. Mix to combine, then gently mix in the roasted vegetables. Season to taste and serve warm or at room temperature.

Grilled corn from the cob and tomato salad

Grilled corn from the cob and tomato salad

Clean

In Canada, nothing says summer quite like local corn-on-the-cob and sweet tomatoes. Here, the two are joined in a simple, Mexican-inspired, colorful salad.

Serves 4

3 **corn cobs**
2 **limes,** 1 juiced, 1 cut into quarters
⅔ cup (150ml) **mayonnaise or plain Greek yogurt**
¼ tsp **cayenne**
4oz (100g) **feta**
6 **large multi-colored tomatoes,** quartered
Sea salt and freshly ground pepper

To serve:
Fresh chives, roughly chopped
Lime wedges

Preheat a grill or grill pan to a high heat.

Soak the corn husks for 20 minutes, then remove the husks and silks. Squeeze the lime juice over the corn. Grill, turning every few minutes, until browned in spots, about 12 minutes. Cool slightly.

Combine the mayonnaise or yogurt, cayenne, feta and juice of 1 lime in a small bowl. When corn is cool enough to handle, cut off the kernels, keeping them intact if possible. Layer the corn with the sliced tomatoes on four plates and sprinkle with chopped chives. Serve with bread, a lime wedge and a dollop of dressing.

Orzo with broad beans, Pecorino and quince

Spring

I think of orzo as a cross between pasta and rice, and it makes a wonderful salad ingredient. Other similar pasta shapes are available, such as puntalette or riso, so use whatever is available. If you're struggling to find any of these, then any small pasta shape or even couscous will be just as delicious.

Serves 6

3½ cups (500g) **orzo**
Scant 1 cup (200g) **fresh podded broad beans**
Scant 1 cup (200g) **fresh** shelled **peas**
4 Tbsp **extra-virgin olive oil**
Juice and zest of 2 **lemons**
1 **garlic clove,** peeled and crushed
4oz (100g) **quince paste,** very finely cubed or chopped
1½ cups (100g) **finely grated Pecorino or Parmesan cheese**
3–4 Tbsp **fresh oregano and/or thyme leaves**
Salt and freshly ground black pepper

Cook the orzo according to the directions on the package until *al dente*. Run under cold water, strain well and set aside.

Cook the broad beans and peas separately until just tender, then refresh them quickly under cold, running water. Remove the grey outer skin from the broad beans by tearing and gently squeezing out the green center.

Combine the olive oil, lemon juice and garlic together and pour over the orzo. Stir in the broad beans, peas, quince paste, cheese and herbs and season generously to taste. Serve at room temperature.

Larry's eggplant and blood orange salad

Impressive

When not rocking out on his guitar, my chef friend Larry Fogg is whipping up stunning culinary creations (while simultaneously playing with his daughters Harriet and Pearl). This is just one from his collection.

Serves 4

1 **small eggplant,** cubed
Salt
3 Tbsp **olive oil**
1 **garlic clove,** peeled and finely chopped
¾-in (2-cm) **piece fresh root ginger,** peeled and finely chopped
2 **blood oranges (see tip below)**
½ cup (100ml) **balsamic vinegar**
4 Tbsp **honey**
Sea salt and freshly ground black pepper to taste
5 cups (200g) **salad leaves**
1 **small red onion,** peeled and finely sliced

Place the cubed eggplant in a colander and sprinkle liberally with salt. Leave for 1 hour, then rinse and pat dry. (This will remove bitter water from the eggplant.)

Heat 1 Tbsp olive oil in a frying pan on a medium-high heat. Add the eggplant, garlic and ginger and sauté until the eggplant is golden. Set aside.

To segment the oranges, slice a small section off the top and bottom and stand the orange on end. Following the contours of the fruit, slice off the skin and pith. Remove sections by slicing along the membranes. Set the segments aside. Squeeze the juice from the membrane into a small saucepan. Add the balsamic vinegar and honey. Place the saucepan over a high heat and bring the contents to a boil. Stir for 1 minute, add the eggplant and simmer gently for 5 minutes. Remove from the heat and leave to infuse for at least 1 hour.

Before serving the salad, use a slotted spoon to remove the eggplant from the balsamic mixture and set aside. Place the saucepan over medium heat and gently simmer the balsamic mixture until reduced to a thick syrup. Whisk in the remaining olive oil and season with salt and pepper.

Arrange the salad leaves on four plates and top with orange segments, onion and eggplant. Drizzle with balsamic vinaigrette and serve.

■ *Look for blood oranges in winter when they are in season. If you can't find them, use ordinary oranges instead.*

Minted rice and pistachio salad

Minted rice and pistachio salad

Jeweled

Pink peppercorns and green pistachio nuts make this salad as easy on the eye as it is on the stomach!

Serves 4

¾ cup (100g) **wild rice**
¾ cup (100g) **basmati rice**
½ cup (125ml) **extra-virgin olive oil**
Zest and juice of 2 **lemons**
2 **garlic cloves,** peeled and crushed
Large handful **fresh mint leaves,**
  finely shredded
Small handful **fresh flat-leaf parsley,**
  finely shredded
½ cup (75g) **dried blueberries**
½ cup (50g) **pistachios,** toasted
½ cup (50g) **slivered almonds,**
  toasted
2 Tbsp **pink peppercorns,** lightly
  crushed
Salt and freshly ground black
  pepper

Cook the two rice varieties separately as directed on the packages. Drain and set aside to cool completely.

In a small bowl combine the oil, lemon zest, juice and garlic. Pour this mixture over the rice, add the herbs and stir until well combined. Stir in the pistachios, half of the almonds and peppercorns and mix well. Season to taste with salt and pepper and serve sprinkled with the remaining toasted almonds.

■ *Cool the rice as quickly as possible by removing from the pan and cooling in a larger dish. As soon as it is cool cover and refrigerate until needed.*

Moroccan couscous and honey-roasted carrot salad

Sweet

This salad is a hearty and nutritious meal in itself. The carrots intensify in flavor when roasted in this way and add a subtle sweet note to the dish.

Serves 4–6

1lb 2oz (500g) **carrots,** peeled and cut on the diagonal about ½in (1-cm) thick
2 Tbsp **olive oil**
2 Tbsp **honey**
¼ tsp **ground cinnamon**
½ tsp **sweet smoked paprika**
14-oz (400-g) **can chickpeas,** drained and rinsed
1 cup (200g) **couscous**
1¼ cups (300ml) **boiling hot vegetable stock**
Juice of 1 **lemon**
4oz (100g) **feta,** cut into cubes
2 Tbsp **sesame seeds,** toasted
8 **large dates (preferably medjool),** stone removed and cut into quarters
Salt and freshly ground black pepper
Small handful **fresh cilantro leaves,** roughly chopped

Preheat oven to 375°F (190°C). Put the carrots, oil, honey, cinnamon and paprika in a large roasting pan and mix well to combine. Roast in the oven for 20–25 minutes, stirring once or twice. Stir in the chickpeas for the last 5 minutes of cooking.

While the carrots are cooking prepare the couscous. Put the couscous in a large bowl and pour in the vegetable stock. Stir well, cover and leave to rest for about 20 minutes or until all the liquid has been absorbed.

Combine the cooked carrots, couscous, lemon juice, feta, sesame seeds and dates and season generously with salt and pepper. Stir in the cilantro just before serving. Serve warm or at room temperature.

Moroccan couscous and honey-roasted carrot salad

New potato salad with homemade herb mayonnaise

New potato salad with homemade herb mayonnaise

Herby

There's no comparison to a homemade mayonnaise. It's a real taste of summer when packed with fresh herbs and is a delicious addition to any barbecue.

Serves 4–6

For the mayonnaise:
1 large egg (see note page 4)
1 egg yolk
1 garlic clove, peeled and crushed
1 Tbsp Dijon mustard
1¼ cups (300ml) mild olive oil or vegetable oil
Juice of 1 lemon
Salt and freshly ground black pepper
4–6 Tbsp chopped fresh mixed herbs (e.g. tarragon, basil, dill, parsley)

For the salad:
1lb 10oz (750g) baby new potatoes, boiled until just tender
1 red onion, peeled and finely diced
1 celery stalk, finely diced (optional)

First make the mayonnaise. Put the eggs in a food processor with the garlic and mustard. Blitz until just combined and smooth. With the motor running, slowly drizzle the oil into the bowl of the processor in a steady stream. You must be very careful here as adding too much oil too quickly will cause the mixture to curdle. When all the oil is in, add the lemon juice then switch the motor off. Season generously to taste. Transfer the mayonnaise to a bowl and stir in the herbs.

Put the prepared potatoes, onion and celery into a bowl. Pour over the desired amount of mayonnaise and gently mix to combine. Any extra mayonnaise can be stored in the refrigerator for up to 1 week.

Roasted red beet, grapefruit and pistachio salad

Magenta

Color and texture come together in this vibrant, wintry salad.

Serves 4

6 **small** (about 4oz/100g total weight)
   red beets, tops removed
1 **pink grapefruit**
2½ cups (100g) **salad greens (bib or**
   butter lettuce is best)
2 Tbsp shelled **pistachios**

For the vinaigrette:
1 **shallot,** peeled and finely chopped
1 **rounded tsp sea salt**
1 **garlic clove,** peeled
1 **rounded tsp Dijon mustard**
1 Tbsp **red wine vinegar**
Freshly ground black pepper
5 Tbsp **extra-virgin olive oil**

Preheat the oven to 400°F (200°C). Wash the red beets and pat dry. Place them in foil and wrap up the ends, scrunching to seal. Place the package in the oven and roast for 1 hour, or until the beets are tender when pierced with a knife. Open the package and allow to cool for 5 minutes. Rub off loose skins and slice each beet into quarters or in half, depending on size.

To segment the grapefruit, slice a small section off the top and bottom. Stand the grapefruit on end and, following the contours of the fruit, slice off the skin and pith. Remove the sections by slicing along the membranes. Set segments aside.

To make the vinaigrette, combine the shallot, sea salt and garlic in a mortar and pestle. Pound together to a purée. Whisk in the mustard, vinegar and pepper. Slowly whisk in the oil.

Arrange the greens on four plates. Place the beets and grapefruit segments on the greens. Drizzle with vinaigrette and sprinkle with pistachios.

Roasted tomato, lentil and chestnut soup

Hearty

To me this soup sums up Tuscan food in a bowl. Rich tomatoes, earthy lentils and creamy chestnuts finished off with aromatic herbs is absolute bliss.

Serves 6

7 (2lb) **ripe plum tomatoes**
3 **garlic cloves,** peeled and crushed
1 Tbsp **superfine sugar**
½ tsp **crushed red pepper**
3 Tbsp **olive oil**
1 **onion,** peeled and finely chopped
2 **bay leaves**
2 **large sprigs fresh thyme**
Scant 1 cup (200g) **green (puy) or brown lentils,** rinsed
2 Tbsp **tomato paste**
4 cups (2L) **hot vegetable stock or water**
1 cup (500g) **vacuum-packed chestnuts,** roughly chopped
Salt and freshly ground black pepper
Large handful **fresh flat-leaf parsley,** roughly chopped

Preheat heat the oven to 375°F (190°C). Roughly chop the tomatoes into about thirds or quarters and place in a roasting pan. Sprinkle over the garlic, sugar and crushed red pepper and drizzle on 2 Tbsp of the oil. Toss gently together then roast in the oven for 40–45 minutes, stirring occasionally.

While the tomatoes are cooking, heat the remaining tablespoon of oil in a large stockpot. Add the onion and sauté until translucent but not browned. Stir in the bay leaves, thyme and lentils and stir to coat in the oil. Add the tomato paste and stock and bring to a boil. Reduce the heat and simmer gently for 30–40 minutes or until the lentils are tender. Add the roasted tomatoes and chestnuts and simmer for a further 10 minutes. Add salt and pepper to taste, stir in the parsley and serve.

Asparagus, frisée and poached quail's eggs with sesame dukkah

Gooey

This dish not only looks fantastic but it tastes fantastic too. It's a great starter that is simple yet impressive. Serve any leftover dukkah with fresh bread and extra-virgin olive oil on the side.

Serves 4

For the sesame dukkah:
1 Tbsp **coriander seeds**
1 Tbsp **cumin seeds**
2 Tbsp **white sesame seeds**
1 tsp **sea salt flakes**
½ tsp **freshly ground black pepper**

20 **asparagus spears,** ends trimmed
¼ stick (30g) **butter,** melted
12 **quail's eggs**
1 **frisée lettuce,** leaves separated

First, make the sesame dukkah. In a dry, nonstick frying pan heat all the ingredients until lightly toasted and aromatic. Remove from the heat and transfer to a mortar. Grind to a coarse powder with the pestle and set aside. Alternatively, you can grind the seeds in a coffee grinder dedicated only to grinding spices.

Bring two pans of water to a boil. In one pan cook the asparagus until just tender then drain. Pour over the melted butter and half of the spice mix and gently mix, then set aside.

Reduce the second pan to a simmer and carefully crack the eggs into the water, one at a time. Continue to barely simmer until the white has set around the yolk. Lift out using a slotted spoon and drain on paper towel.

Put five asparagus spears on each plate. Top with a small handful of frisée lettuce leaves and three quail's eggs. Sprinkle over the remaining spice mix evenly among the plates and pour over any remaining butter. Serve immediately.

Asparagus, frisée and poached quail's eggs with sesame dukkah

Bulgar wheat salad with green beans and sumac

Zingy

Sumac has a sour, lemony flavor that was traditionally used as a substitute for lemons. It is commonly used in Lebanese, Syrian and Turkish cooking. If you can't get hold of sumac, use the same amount of grated lemon zest.

Serves 4–6

1½ cups (250g) **bulgar wheat**
1½ cups (200g) **fine green beans,** trimmed
4 Tbsp **extra-virgin olive oil**
Juice of 1 **lemon**
1½ tsp **sumac**
2 **garlic cloves,** peeled and crushed
Large handful **mint,** roughly chopped
Large handful **parsley,** roughly chopped
9oz (250g) **bocconcini (mini mozzarella balls)**
12 (250g) **cherry tomatoes,** halved
3 **green onions,** trimmed and finely sliced
Salt and freshly ground black pepper

Put the bulgar wheat in a large bowl and generously cover with boiling water. Stir to combine, then cover and leave for 30 minutes.

While the bulgar wheat is soaking, bring a large pan of water to a boil and blanch the beans until just tender but cooked. Drain and run under cold running water. Cut the beans into about 1-in (2.5-cm) pieces, then set aside.

Drain off any excess water from the bulgar wheat and return to the bowl with the green beans. In a separate bowl combine the olive oil, lemon juice, sumac, garlic and herbs. Pour this mixture over the bulgar wheat and mix well. Stir in the bocconcini, cherry tomatoes and green onions and season generously with salt and pepper.

Caponata

Warming

This sweet and sour eggplant salad is a Sicilian specialty. It is rich in both flavor and texture, and makes a great addition to any antipasto platter; also delicious served with fried halloumi.

Serves 4–6

4 Tbsp **olive oil**
1 **onion, peeled and finely chopped**
2 **garlic cloves, peeled and crushed**
1½lb (750g) **eggplants, cut into**
  ¾-in (2-cm) cubes
2 **celery stalks, finely sliced**
1 Tbsp **tomato paste**
2 Tbsp **capers, rinsed**
14-fl oz (400-ml) **can chopped**
  tomatoes
6 Tbsp (75ml) **red wine vinegar**
  or cider vinegar
2 Tbsp **sugar**
Salt and freshly ground black
  pepper
Small bunch fresh flat-leaf parsley,
  chopped
½ cup (50g) **almonds or pine nuts,**
  sliced and toasted

Heat the oil in a large, heavy-based saucepan and add the onion. Sauté over medium heat until translucent and cooked, but not brown. Add the garlic and stir to combine. Then add the eggplant and celery and stir until coated in oil.

Continue to sweat over a low heat until the eggplant begins to soften. Stir in the tomato paste then add the capers, tomatoes, vinegar and sugar and mix well to combine. Bring to a gentle simmer and cook for a further 15–20 minutes, stirring occasionally.

Season to taste, then stir in the almonds and parsley and serve warm or at room temperature.

Asparagus, pea and toasted quinoa salad

Vibrant

Sometimes it is difficult to eat something that looks as pretty as this salad, but you can always make it again tomorrow.

Serves 4

4 Tbsp **freshly-cooked quinoa**
1 **bunch asparagus,** ends trimmed,
    sliced into 1½-in (4-cm) lengths
1¼ cups (200g) **sugar snap peas**
Scant 1 cup (100g) **frozen peas,**
    thawed or **shelled edamame**
1 **head raddichio,** leaves separated
½ cup (10g) **fresh mint leaves,**
    finely sliced

For the vinaigrette:

4 Tbsp **extra-virgin olive oil**
½ tsp **maple syrup**
2 Tbsp **balsamic vinegar**
1 tsp **Dijon mustard**
Pinch **brown sugar**
Pinch **sea salt flakes**
Pinch **cracked black pepper**

Heat a small frying pan over medium-high heat. Spoon the quinoa into the dry, hot pan and toast, stirring constantly to separate, until the grains are dry and toasted. Transfer to a bowl and cool.

Bring a saucepan of water to a boil. Blanch the asparagus, sugar snap peas and peas together for just 1–2 minutes, until bright and almost tender. Drain through a sieve and drench with plenty of cold water. Set aside.

Whisk the vinaigrette ingredients together in a large bowl. Add the blanched vegetables and toss to coat.

Arrange the raddichio on four plates. Top with the blanched vegetables, toasted quinoa and fresh mint.

■ *This is the perfect recipe when you have leftover quinoa in the fridge. We found red quinoa for this recipe but any kind will do.*

Asparagus, pea and toasted quinoa salad

Artichoke and herb tartlets

Impressive

It requires a little bit of patience to prepare the artichoke, but I can assure you it is worth it. You will be rewarded with every mouthful.

Makes 6

6 large globe artichokes
Juice of 1 lemon
1 cup (250ml) white wine
1 cup (250ml) tomato passata
   (drained tomatoes)
3 bay leaves
2 Tbsp olive oil
1 small onion, peeled and finely
   chopped
1 garlic clove, peeled and crushed
3 Tbsp chopped fresh mixed herbs
   (e.g. mint, parsley, chives,
   oregano, rosemary)
1 egg, beaten
1 Tbsp milk
3oz (75g) goat's cheese
Salad leaves to serve

Prepare the artichokes. Using a sharp knife, slice the stems off at the base and snap away the leaves one by one and discard them. Trim the base of any dark green bits and use a small spoon to dig out the furry choke. Rub the artichokes immediately with lemon juice as you go and set the bases aside in a small saucepan. When all of the artichokes are prepared, add the wine, passata, bay leaves and 1 Tbsp of oil to the pan. Bring to a boil, then reduce the heat and simmer for 20–25 minutes or until the bases are tender. Leave to cool in the liquid.

Preheat the oven to 400°F (200°C). Prepare the filling by heating the remaining oil in a nonstick frying pan. Cook the onion for 3–4 minutes or until soft and translucent. Stir in the garlic and cook for a further minute. Remove from the heat and stir in the herbs. Put the onion, garlic and herb mixture into a bowl and stir in the egg and milk.

To assemble the tartlets, remove the artichoke bases from the liquid and pat dry. Place them flat on a baking tray. Spoon the onion, garlic and herb mixture into the bases, then crumble over the goat's cheese. Bake in the oven for 10–15 minutes or until the filling is set and lightly golden.

Serve warm or cold, either whole or cut in quarters, on a bed of salad leaves

Spanish roasted vegetables

Rustic

The Spanish name for this dish is escalivada. These Catalan-roasted vegetables are delicious served warm or cold, on their own or as an accompaniment to other dishes, or simply soaked up by warm homemade bread.

Serves 4–6

2 **red peppers**
2 **medium eggplants**
3 **small red onions**
3 **garlic cloves**
4 **tomatoes**
3–4 Tbsp **extra-virgin olive oil**
Juice of 1 **lemon**
Salt and freshly ground black
    pepper to taste

Preheat the oven to 400°F (200°C). Wash all the vegetables well and dry using paper towel. Lay them in a single layer on a large baking tray and add enough olive oil to lightly coat them. Season with salt and pepper and roast in the oven for 20–25 minutes or until soft but still holding their shape, the skins loosened and slightly charred in parts. Remove from the oven and leave to cool.

When cool enough to handle, peel the skins off the vegetables. Remove the cores and seeds and chop into large chunks. Combine the lemon juice and 1 Tbsp of olive oil in a separate bowl and season well. Pour over the vegetables and serve warm or cold.

Chèvre-stuffed zucchini flowers

Delicate

Zucchini flowers have a short season so keep your eye out for them. Once harvested, shake the flower gently to remove any dirt, then trim the stems (i.e. the young zucchini) to about 2in (5cm) in length.

Serves 6

For the batter:
1¼ cups (150g) **all-purpose flour**
¼ tsp **salt**
1 **egg**
Approx. 1¼ cups (300ml) **lager**

For the filling:
10oz (300g) **creamy cheese – chèvre, mascarpone or ricotta, or all three**
1 **egg**, beaten
2 Tbsp **fresh mint leaves**, finely chopped
Sea salt and freshly ground black pepper to taste
12 **large organic zucchini flowers**

Vegetable oil for deep-frying
2½ cups (100g) **arugula**
6 Tbsp **extra-virgin olive oil**
3 Tbsp **balsamic vinegar**
Handful **raspberries**

Make the batter. Combine the flour and salt in a large bowl. Whisk in the egg, then slowly add enough lager to make a thick but still runny batter. Set aside for 1 hour before using.

To make the filling, combine the cheese, beaten egg, chopped mint, salt and pepper in a small bowl. Gently open the zucchini flowers and place approximately 1 rounded tsp of filling into each flower. Gently twist the flower petals to contain the filling.

Heat the oil in a large, deep saucepan until a drop of batter sizzles when dropped into the oil. Holding the flowers by their stem, dip them into the batter, one at a time, then carefully put into the hot oil. Deep-fry for a few minutes, turning once, until crisp and golden. Drain on paper towel and sprinkle with salt.

Arrange arugula on serving plates. Whisk the olive oil and balsamic vinegar together, and season to taste. Add the raspberries and crush slightly. Arrange the flowers over the arugula and drizzle with the vinaigrette.

Chèvre-stuffed zucchini flowers

Halloumi with grilled zucchini and mint

Fresh

Halloumi is a white ewes' milk cheese with a firm texture, which makes it excellent for grilling and frying. Try a selection of other grilled vegetables, such as eggplant and peppers, for a different approach.

Serves 4

2–3 (about 10–12oz/300–350g) **zucchini**
3½ Tbsp **extra-virgin olive oil**
Small handful **fresh mint**, roughly chopped
Juice of 1 **lemon**
Salt and freshly ground black pepper
10oz (300g) **halloumi**, sliced ½in (1cm) thick

Using a very sharp knife or mandoline, finely slice the zucchini lengthwise to about $\frac{1}{16}$–$\frac{1}{8}$in (2–3mm) thick. Preheat a grill pan to very hot. Using 1 Tbsp of the oil, lightly brush the zucchini slices and grill for about 2 minutes on each side until grill marks appear and the zucchini is cooked. Transfer to a bowl and continue until all the zucchini is cooked.

Combine 2 Tbsp of the oil with the mint and lemon juice and pour over the warm zucchini. Stir to combine and season generously with salt and pepper.

Heat the remaining oil in a large nonstick frying pan and cook the halloumi for 2–3 minutes, turning once, until golden and warmed through.

Serve 2–3 slices of halloumi on each plate, with a quarter of the zucchini on top and drizzled with any extra dressing and a sprinkle of black pepper. Serve immediately.

Halloumi with grilled zucchini and mint

Green, white and red ribollita
Fresh

Pronounced "ree-bo-LEE-tah" and meaning "twice boiled," this is a Tuscan soup originally made from leftover minestrone. There are countless variations and permutations of this classic, but this version is packed with all good things – green, white and red.

Serves 6

2 Tbsp **olive oil**
1 **large onion,** peeled and roughly chopped
2 **celery stalks,** leaves and stalks chopped
3 **garlic cloves,** peeled and minced
1 bunch **chard,** roughly chopped
5 cups (200g) **spinach,** roughly chopped
Large handful **fresh mint,** roughly chopped
Large handful **fresh basil,** roughly chopped
19-oz (525-g) **can tomatoes**
Parmesan rind (see note)
20-oz (540-g) **can cannellini beans (or white kidney beans),** drained and rinsed

To serve:
Grated Parmesan cheese
Extra-virgin olive oil
Crusty bread

Heat the olive oil over medium heat in a large soup pot. Add the onions and celery and sauté for 5 minutes, until the onions begin to soften. Add the garlic, chard and spinach and cook for a further 10 minutes, until everything is flavorful and soft. Add half the mint, half the basil, the tomatoes and Parmesan rind. Stir well. Stir in the beans and enough water to cover. Cook for a further 15 minutes.

Before serving, discard the Parmesan rind and stir in the remaining herbs. Divide the soup among six warmed bowls and sprinkle on some Parmesan cheese and a drizzle of extra-virgin olive oil.

Serve with crusty bread.

■ *Parmesan rinds will keep tightly wrapped in the fridge for up to 2 months, or frozen for 6 months. They are an excellent flavor addition to soups and stews, whatever their size.*

Miso and sea vegetable soup

Cleansing

Add some cubed tofu to this dish or a handful of sprouts for a heartier alternative.

Serves 4

1 tsp (5g) **sea vegetables (e.g. wakame, dulse, agar, hijiki, kelp)**
4½ cups (1L) **cold water**
2 Tbsp **miso paste**
2–3 **Savoy cabbage leaves,** shredded
3oz (70g) **rice noodles,** cooked, drained and rinsed in cold water

Rinse the sea vegetables to remove excess salt, then soak them in warm (boiled) water for 5 minutes. Bring 4½ cups (1L) cold water to a boil in a large saucepan. Drain the sea vegetables then add to the pan.

Spoon about 2 Tbsp of water from the pan into a small bowl. Stir in the miso paste until dissolved, then add to the pan with the cabbage and noodles. Reduce to a simmer and cook for a further 4–5 minutes. Serve immediately.

Vietnamese spring rolls, unrolled

Balance

Pippa and I couldn't write a vegetable book without including one of our favorite gastronomic delights, the Vietnamese spring roll. We all know, however, that the rice paper step can be tricky, so why not just eat the insides?

Serves 6

4oz (100g) **rice noodles,** freshly cooked and rinsed
1 **mango,** peeled and flesh chopped into matchsticks
2 **carrots,** peeled and chopped into matchsticks
1 **small red onion,** peeled and finely sliced
2-in (5-cm) **piece cucumber,** chopped into matchsticks
2 Tbsp chopped **fresh mint**
2 Tbsp chopped **fresh cilantro**
2 Tbsp chopped **fresh chives**

For the dressing:
1 **birdseye chili,** halved, seeds removed, flesh shredded
2 Tbsp **sweet chili sauce**
2 Tbsp **palm or brown sugar**
4 Tbsp **fish sauce**
2/3 cup (150ml) **fresh lime juice**

To serve:
Handful **of toasted peanuts or cashews,** roughly chopped
1 Tbsp (10g) **sprouts**

Combine the dressing ingredients in a small bowl. Whisk well and set aside.

Put the rice noodles in a serving bowl. Top with the remaining ingredients, except the peanuts, pour over the dressing and gently toss. Garnish with peanuts or cashews and sprouts.

Vietnamese spring rolls, unrolled

Snacks and small dishes

Pippa and I are on a two-woman mission. We would like each and every one of you to taste vegetables, again, for the very first time. Let's start with the ubiquitous vegetable tray. We prefer to call it *crudités*. Somehow its French name elevates mere carrot sticks to a higher, holier level. "Vegetable tray" denotes wilted carrots, huge, unfriendly florets of broccoli, dry cauliflower and limp celery, all fanned around a tub of store-bought dip. It's uninspiring; it's boring; it's sad.

But cheer up. *Larousse Gastronomique*, the Bible for all things culinary, states that *crudités* include chopped fennel, celeriac, tiny red radishes, grape tomatoes, broad beans, snow peas and even slices of banana sprinkled with lemon, for goodness sakes! We would like to extend the definition even further into the realm of the roasted vegetable.

Of course snacks and small dishes are not limited to the – albeit new and improved – *crudité* platter. In the following pages you will find everything from cornbread-crusted zucchini to tenderstem broccoli tempura with soy chili dipping sauce.

Armenian baba ghanoush

Armenian baba ghanoush

Smoky

This version of the ubiquitous smoky eggplant dish includes peppers and onions, roasted until they are meltingly soft. The finished dish is a colorful and rustic appetizer. Serve with toasted pitas or flatbreads.

Serves 4

1 **large** (1lb 10oz/750g) **eggplant**
1 **large red pepper**, left whole
1 **large onion**, peeled and quartered
6 **garlic cloves**, peeled
4 Tbsp **olive oil**
1½ tsp **cumin seeds**
1½ tsp **coriander seeds**
1½ tsp **sea salt**
Juice of 1 **lemon**
Handful fresh cilantro, roughly chopped (optional)

Preheat the oven to 400°F (200°C). Prick the eggplant all over with a fork and place directly on a rack in the oven. Place the remaining vegetables and garlic in a small roasting pan. Toss with 2 Tbsp of the olive oil and roast for 45 minutes, stirring every 15 minutes, until the eggplant is very soft and deflated, the pepper is charred and the onions and garlic are golden and slightly charred. Allow to cool.

Meanwhile, put the cumin and coriander seeds in a small frying pan and place over a medium heat. Cook until the seeds are fragrant and begin to pop, about 2 minutes. Transfer the seeds to a pestle and mortar, along with the sea salt, and pound until everything is combined and cracked, but not perfectly smooth.

When the vegetables are cool enough to handle, cut open the eggplant, scoop out the flesh and place on a cutting board. Peel, core and deseed the pepper and place on the cutting board with the onion and garlic. Roughly chop everything and transfer to a serving dish. Toss with spices, lemon juice and remaining olive oil. Garnish with cilantro (optional) and serve.

Baby gem with sweetcorn and avocado salsa

Crisp

This salsa is so colorful and vibrant and, if made in larger quantities, works equally well as a salad in its own right. Try serving with corn chips.

Makes 24

24 **baby gem leaves or other small lettuce leaves**

For the salsa:
2 **corn cobs**
2 **green onions**, finely sliced
1 **red chili**, seeded and finely chopped
1 **large avocado**, stoned, peeled and diced into ½-in (1-cm) cubes
Juice of 2 **limes**
Small handful fresh cilantro, roughly chopped
Sea salt and freshly ground black pepper

Cilantro leaves, roughly chopped, to garnish (optional)
Lime wedges, to serve (optional)

Carefully wash and separate the lettuce leaves and dab with paper towel to remove excess water. Place, cut side up, on a serving platter and set aside.

To prepare the salsa, bring a large saucepan of water to a boil. Add the corn cobs and blanch for 3–4 minutes, drain, then run under cold water until cool enough to handle. Using a large sharp knife, slice the kernels from the cob, as close to the cob as possible. Put the corn in a large mixing bowl and add the remaining ingredients. Combine gently and season accordingly.

Place a heaped Tbsp of the salsa into each baby gem leaf and garnish with cilantro leaves, if liked. Serve immediately with lime wedges.

Baby gem with sweetcorn and avocado salsa

Cornbread-crusted zucchini with mango salsa

Cornbread-crusted zucchini with mango salsa

Punchy

One of the many wonderful dishes one can make with the summer's zucchini crop. The salsa (see page 162), with a hint of mango sweetness and a hit of chili, is nice and spicy. It can be tamed, however, depending on taste.

Serves 4

4 Tbsp **all-purpose flour**
½ cup (50g) **fine cornmeal (semolina)**
¼ tsp **flaky sea salt**
Pinch **cayenne pepper**
1 **egg,** beaten
3 **zucchini (approx. 4in/10cm long)**
4 Tbsp **vegetable oil**
Mango salsa (see page 162)

Combine the flour, cornmeal, salt and cayenne in a shallow bowl. Beat the egg in a separate shallow bowl and set aside.

Cut the zucchini into ½-in (1-cm) slices. Heat the oil in a large, heavy-based saucepan to 375°F (190°C), or until a small cube of bread turns golden in about 30 seconds. Dip the zucchini slices in the beaten egg, then the semolina mixture and fry for 2–3 minutes on each side until golden. Drain on paper towel. Serve warm with the salsa.

Endive leaves with pear, blue cheese and hazelnut
Classic

This is a classic combination that works equally well as a salad as it does a canapé. Baby gem leaves also work well as an alternative to endive.

Makes 12

1 cup (250ml) **sweet dessert wine**
1 cup (250ml) **cold water**
¾ cup (150g) **superfine sugar**
Juice of 1 **lemon**
2 **star anise**
1 **cinnamon stick,** broken in half
2 **pears, peeled**
12 **endive leaves**
4oz (100g) **blue cheese**
½ cup (25g) **hazelnuts,** roughly but quite finely chopped

Put the wine, water, sugar, lemon, star anise and cinnamon in a medium saucepan and bring to a boil. Reduce the heat and simmer for 5 minutes. Add the pears and continue to simmer for 15–18 minutes or until a sharp knife can easily be inserted into the center of the pears. Remove from the heat and allow the pears to cool in the poaching liquid. When cold, slice each pear into about 12 slices and set aside.

Place the endive leaves on a serving plate and put about two slices of pear into each. Top with a slice of blue cheese and a sprinkling of hazelnuts and serve immediately.

Creamy white bean and fresh herb dip

Clean

A quick, flavorful dip that is a refreshing alternative to hummus. Serve with raw vegetables, lavash (a thin Armenian cracker bread) or crispy pitas.

Makes 1¾ cups (400ml)

19-oz (525-g) **can cannelini, navy or white kidney beans,** drained and rinsed
2 Tbsp freshly-squeezed **lemon juice**
2 Tbsp **extra-virgin olive oil**
1–2 **garlic cloves,** peeled
1 tsp **ground cumin**
2 Tbsp chopped **fresh herbs** (thyme, mint, rosemary)
2 Tbsp **Greek yogurt**
Zest of ½ **lemon**
Sea salt and freshly ground black pepper to taste

Combine the first five ingredients in a food processor and pulse until almost smooth. Add the remaining ingredients and pulse once or twice until combined. Spoon into a bowl to serve.

■ *This dip can be made up to a day in advance and refrigerated. Bring to room temperature before serving.*

Crudités with red pepper, cashew and feta dip

Vibrant

This dip also makes a fantastic pasta sauce when stirred through some hot penne with a few arugula leaves on top.

Makes about 10fl oz (300ml)

2 **red bell peppers**
½ cup (100ml) **extra-virgin olive oil**
¾ cup (75g) **unsalted cashew nuts**
4oz (100g) **feta cheese**
⅛–¼ tsp **cayenne pepper**
Crudités to serve (e.g. celery, carrots, cucumber, snow peas, broccoli, cauliflower, radishes)

Preheat the oven to 350°F (180°C). Lightly rub the peppers with a little of the olive oil, place on a baking tray and roast in the preheated oven for about 30–40 minutes, turning once or twice, until slightly charred and softened. Remove from the oven, cover with foil and allow to cool.

When the peppers are cool enough to handle, peel off the skin and remove the core and seeds. Place the flesh in a food processor. Add the cashew nuts, feta and cayenne pepper and blitz to combine. With the motor running, gradually add the remaining olive oil until well combined.

To prepare the crudités, wash and trim the vegetables, removing any unwanted stalks, leaves or outer layers. Chop into smaller pieces if necessary so they can be easily picked up and eaten. Serve the dip with the crudités on the side.

Crudités with red pepper, cashew and feta dip

Polenta with fig and red onion relish and goat's cheese

Polenta with fig and red onion relish and goat's cheese

Rustic

If you don't like any waste, cut the polenta into small triangles instead of circles. This will avoid any scraps of polenta being discarded.

Make 20–24

4½ cups (1L) **vegetable stock or cold water**
Generous 2 cups (250g) **cornmeal or coarse maize**
1 tsp **salt**
1 tsp **dried oregano, optional**
1–2 Tbsp **olive oil**
Fig and red onion relish
(see page 160)
3oz (75g) **goat's cheese**
Freshly ground black pepper

Bring the stock to a boil in a large saucepan. Combine the cornmeal, salt and oregano then quickly whisk them into the boiling liquid. As soon as it comes to a boil again, reduce the heat and exchange your whisk for a wooden spoon. Simmer the cornmeal, stirring regularly, for about 10 minutes or until thick and porridge-like. Be careful as it will splatter volcanically. Line a 12 x 8-in (30 x 20-cm) baking tray with parchment paper and quickly spread the cornmeal evenly over it so it is no more than ½in (1cm) thick. Use a dampened pallet knife to spread it evenly and smooth the surface. Leave the cornmeal to cool and set for at least 30 minutes. (The recipe can be prepared up to two days in advance to this stage.)

Using a 1¼–1½-in (3–4-cm) round cutter, cut out rounds from the cornmeal. Brush both sides of each round with olive oil and cook the wedges either under a hot broiler or in a grill pan, turning once, until golden and crisp. Alternatively heat 1 Tbsp oil in a non-stick frying pan and fry both sides until golden brown, about 2 minutes each side.

Place the polenta rounds on a serving platter. Put a dollop of Fig and red onion relish on each round and top with a small cube or crumble of goat's cheese and a grind of black pepper. Serve warm or at room temperature.

Potato farls with dill butter

Warming

Potato farls are Northern Irish breads traditionally served with breakfast, but I recently ate farls – torn apart, I should say – between sips of Black Velvet on St. Patrick's Day. Needless to say, they're delicious – day or night.

Serves 6

For the butter:
6 Tbsp **butter, at room temperature**
2 Tbsp **chopped fresh dill**
Freshly ground black pepper

For the farls:
2 medium (350g) **potatoes, peeled and halved**
1 **egg, whisked**
¼ cup (75ml) **buttermilk**
½ tsp **salt**
½ tsp **baking powder**
1¼ cups (150g) **all-purpose flour, plus more for dusting**

Mix the butter and dill in a small bowl. Season with pepper. Spoon into a small serving bowl, cover and chill for up to two days. Bring to room temperature before serving.

Put the potatoes in a large saucepan with enough water to cover. Bring to a boil then simmer, covered, for 10–15 minutes or until tender. Drain well and return to the saucepan. Mash the potatoes, add the egg and buttermilk and stir until mixture is a thin purée. Add the salt and baking powder, stir well, then add enough flour to form a very soft dough.

Gather the mixture into a ball and turn on to a lightly-floured surface. Knead lightly until smooth. Shape the dough into a 7¾-in (20-cm) circle. Heat a large nonstick frying pan over medium-high heat. Place the dough in the pan and cover. Cook for 3 minutes, or until bottom side is golden brown. Turn, cover and cook for a further 3 minutes or until both sides are golden brown and the bread is cooked through.

Serve the bread warm with dill butter.

Potato farls with dill butter

Plantain fritters with sweet chili sauce

Tropical

Plantains, to be precise, are a fruit, not a vegetable. But they are such a savory fruit, usually prepared in a savory way. So despite the title of this book, they've made the cut. In this recipe this Caribbean "vegetable" is given a Southeast Asian twist. There are many good bottled sweet chili sauces available in Asian food shops, but why not make it yourself?

Makes 15 fritters

3 **ripe plantains** (total weight of about 1lb 10oz/750g) **(see tip below)**
½ tsp **baking powder**
1 **small garlic clove,** peeled and minced
4 **green onions, white and pale green part only,** finely chopped
1 tsp **salt**
¼ tsp **cracked pepper**
½ cup (10g) **fresh cilantro,** roughly chopped
4 Tbsp **vegetable, seed or nut oil**
Sweet Thai chili sauce (see page 165) or crème fraîche and chopped fresh cilantro to serve

Chop each plantain, with the skin on, into three equal pieces. Place in a large pot of boiling water and boil until tender – the time this takes will depend on the ripeness of the plantains. Drain and leave to cool.

Peel the cooked plantains and place the flesh in a mixing bowl. Mash until smooth. Add the remaining ingredients, except for the oil and serving ingredients, to the mashed plantains and stir well. Form into rounds, about 2½ x ½in (5 x 1.5cm). The fritters can be wrapped and refrigerated for up to 6 hours at this stage.

When ready to cook, bring the fritters to room temperature. Heat the oil in a large frying pan until very hot. Fry the fritters until golden brown, 3–4 minutes per side. Drain on paper towel.

Serve warm with Sweet chili sauce or a dollop of crème fraîche and a sprinkling of cilantro.

■ *In our experience, plantains will not ripen if purchased when green, so look for yellow plantains with as many black markings as possible. The riper, the better.*

Spanish omelette with Moroccan tomato jam

Simplicity

Spanish omelette is one of my favorite Spanish tapas. I've spiced it up here with Moroccan tomato jam.

Serves 6–8

¼ cup (75ml) **extra-virgin olive oil**
About 4–6 medium (1kg/2¼lb) **floury potatoes (such as King Edward or Russet),** peeled, quartered and thinly sliced
1 **medium onion,** peeled and thinly sliced
6 **large eggs**
Salt
Moroccan tomato jam
  (see page 161)

Heat the oil in a large nonstick frying pan. Sauté the sliced potatoes and onions, gently lifting and turning until cooked but not brown. This will take about 15–20 minutes.

Beat the eggs together in a large mixing bowl and season well with salt. When the potatoes and onions are cooked, transfer them, using a slotted spoon, into the egg mixture, leaving as much oil behind as possible. Strain the oil from the pan and reserve.

Heat 1 Tbsp of the reserved oil in the same pan over medium heat. Add the egg and potato mixture, spreading the potatoes evenly in the pan. Cook the omelet, covered, for about 5–6 minutes, gently shaking the pan so it doesn't stick. Slide a long spatula underneath the omelet. Place a large plate over it and quickly turn it out on to the plate. Add another Tbsp of oil to the pan and quickly slide the omelet back in. Tuck the sides in with a fork and cook for a further 3–4 minutes or until cooked. Remove from the pan and leave to cool to room temperature.

Cut into cubes or wedges to serve, topped with Moroccan tomato jam.

Root vegetable crisps with garlic aioli

Textured

Root vegetable crisps can be very expensive when you buy them from your local deli, so why not try making your own? They're cheap as chips to make and far more satisfying. A pinch of saffron powder or saffron strands in the aioli will add extra color and flavor too.

Serves 4–6

1lb 2oz (500g) **root vegetables** (e.g. parsnip, celeriac, carrot, red beet, sweet potato)
2¼ cups (500ml) **peanut oil**
Fine salt

For the aioli:
4–6 **garlic cloves,** finely chopped
½ tsp **fine salt**
2 **large egg yolks (see note page 4)**
¼ cup (60ml) **extra-virgin olive oil**
1/2 cup (100–125ml) **vegetable oil** (such as sunflower or canola)

Peel the vegetables and slice them, as thinly as possible, into long strips or rounds using a very sharp knife or mandoline cutter. Heat the oil in a large heavy-based saucepan to 375°F (190°C) or until a small cube of bread turns golden in about 30 seconds. Deep-fry the vegetables in batches (a small handful at a time), for about 2 minutes each batch. As they cook the vegetable slices will curl up and turn golden. Remove with a slotted spoon and drain on paper towel, sprinkling with salt while still warm.

Now prepare the aioli. Put the garlic cloves and salt in a mortar and grind to a paste using the pestle. (Alternatively use a large, heavy knife and mash and chop the garlic to a paste.) Put the garlic in a small food processor or blender and add the egg yolks. Mix well, then slowly add the combined oils, small quantities at a time, until the mixture is very thick. Use immediately or transfer to an airtight container or jar and refrigerate for up to two days.

Serve the crisps with the aioli on the side for dipping.

■ *The root vegetable crisps can be made in advance and stored in an airtight container.*

Root vegetable crisps with garlic aioli

Roasted mixed peppers on toast

Roasted mixed peppers on toast

Reminiscent

Another lovely starter from my cousin Betsy's Tuscan kitchen. Recently she and George sold their home in Stielle, which was sad for the whole family. But that's what recipes are for – they make memories last forever.

Serves 6

2 **large red peppers**
2 **large yellow peppers**
4 Tbsp **olive oil**
2 **garlic cloves, peeled,** 1 minced, **the other** cut in half
2 Tbsp **chopped parsley**
Sea salt and freshly ground black pepper to taste
2oz (50g) **anchovy fillets,** chopped or 2 Tbsp **capers,** rinsed
1 **baguette,** sliced diagonally

Preheat a grill, grill pan or broiler to a high heat. Grill the peppers whole until charred all over. Allow to cool. When cool enough to handle, peel the peppers, remove core and seeds and cut the flesh into thin strips.

Combine the olive oil, minced garlic, parsley, salt and pepper in a bowl. Add the pepper strips and stir to combine. Leave to marinate at room temperature for 1 hour, or in the refrigerator for up to one day. Bring to room temperature before serving.

Preheat the oven to 350°F (180°C). Slice the baguette into 1¼-in (3-cm) thick slices. Arrange on a baking tray and toast for 2 minutes per side until slightly toasted. Rub the cut garlic over the toast. Top with the marinated pepper strips and serve.

■ *This recipe, minus the baguette stage, can be made up to a day in advance. Wrap and store in the refrigerator, but bring to room termperature before spooning over the toasted baguette.*

Spinach, saffron and feta risotto cakes

Substantial

These are delicious served with a spoonful of Moroccan tomato jam (see page 161) or any other chutney on top. Make larger, muffin-size cakes if you prefer, for a great picnic idea.

Makes 50

2 Tbsp **extra-virgin olive oil**
1 **small onion,** peeled and finely chopped
1 **garlic clove,** peeled and crushed
1 **pinch saffron strands**
¾ cup (75g) **risotto rice**
1¼ cups (300ml) **hot vegetable stock**
3 **eggs,** lightly beaten
½ cup (75g) **mascarpone**
2½ cups (100g) **spinach,** blanched and chopped
4oz (100g) **feta,** crumbled
Salt and freshly ground black pepper
1 Tbsp **oil**

First make the risotto. This can be done up to a day in advance. Heat the extra-virgin olive oil in a large, nonstick frying pan and sauté the onion and garlic until soft and translucent but not browned. Stir in the saffron and risotto rice and cook, stirring, for a further 2 minutes. Add the hot vegetable stock, cover and leave to simmer for about 12–15 minutes or until all the liquid has been absorbed and the rice is cooked. Transfer to a bowl and allow to cool.

When you are ready to make the cakes, preheat the oven to 400°F (200°C). Lightly beat the eggs with the mascarpone and add to the risotto mixture along with the spinach and feta. Season generously with salt and pepper and mix well to combine.

Lightly brush the insides of two or four mini muffin trays and place 1 Tbsp of mixture into each case. Cook in the preheated oven for 15–18 minutes or until lightly golden brown and cooked. Cook in batches until all of the mixture has been used. Transfer the cooked cakes to a cooling rack and serve warm or at room temperature.

■ *Cool the risotto as quickly as possible by removing from the pan and cooling in a larger dish. As soon as it is cool it must be covered and refrigerated until needed. Do not reheat the risotto cakes.*

Chickpea quesadillas with pepita pesto

Punchy

Pepitas, an ingredient used primarily in Mexican cooking, are green pumpkin seeds. They add a lovely, nutty taste and texture to pesto. It's the perfect accompaniment to these rich and tasty quesadillas.

Makes 32 wedges

For the pesto:
1½ cups (100g) **pepitas**
2½ cups (50g) **fresh cilantro**
2½ cups (50g) **fresh parsley**
3 **garlic cloves**
1 **fat chipotle chili in adobo sauce** (see tip below)
½ tsp **sea salt**
¼ tsp **freshly ground black pepper**
3 Tbsp **olive oil, plus more for storage**

8 **small (6¾-in/17-cm diameter) flour tortillas**
19-oz (525-g) **can chickpeas**, drained and rinsed
4oz (100g) **soft goat's cheese**
3 Tbsp finely chopped **red onion**
1 tsp **olive oil**

Preheat the oven to 350°F (180°C). Spread the pepitas over a baking tray and toast in the center of the oven, stirring occasionally, for 15 minutes or until golden. Cool slightly.

Transfer the toasted pepitas and remaining pesto ingredients (except for the oil) to a blender or food processor and blend until roughly chopped. Add the oil and blend for a few more seconds until almost smooth. Set aside.

Lay four tortillas on a work surface. Put the chickpeas in a metal bowl and mash roughly with a fork. Add the goat's cheese and red onion and stir until combined. Divide the mixture between the tortillas. Spread 1 Tbsp of pesto on each tortilla, then top with the remaining tortillas to make four "sandwiches." Heat the olive oil in a frying pan over medium heat. Fry the quesadillas, one at a time, on each side until both sides are golden. Transfer to a cutting board and cut each quesadilla into eight wedges.

Top any remaining pesto with a Tbsp of oil, cover and refrigerate for up to three days.

■ *Chipotle chilies are sold in cans, packed in adobo sauce. The remainder of the can may be used to flavor many dishes, from Black bean and vegetable chili (see page 50) to vegetable burgers. Whatever isn't used can be stored in an airtight container for 4–6 weeks.*

Tenderstem broccoli tempura with soy chili dipping sauce

Clean

Try a selection of other vegetables for a bit of variety, such as asparagus, carrots and red peppers. The batter is so light you'll be able to eat more than you think.

Serves 4–6

Dipping sauce:
3 Tbsp **Mirin**
3 Tbsp **white wine vinegar**
4 Tbsp **superfine sugar**
¼–½ tsp **crushed red pepper**
1 Tbsp **soy sauce**
1 Tbsp **sesame seeds**
Juice of 1 **lemon**

1¼ cups (150g) **all-purpose flour**
¼ cup (50g) **cornstarch**
1 cup (250ml) **cold water**
Pinch **salt**
½ tsp **white wine vinegar**
1 **large egg yolk**
14oz (400g) **tenderstem broccoli,**
    long stems trimmed
Vegetable oil, for deep-frying

Put all the sauce ingredients except the sesame seeds and lemon juice in a small saucepan and bring to a boil. Reduce the heat and simmer for 5 minutes. Allow to cool while you prepare the batter. Stir in the lemon juice and sprinkle with sesame seeds just before serving.

Put the flour and cornstarch in a large mixing bowl. Slowly whisk in the water to form a smooth paste. Add the salt, vinegar and egg yolk and mix to combine. Refrigerate the batter for at least 15 minutes or up to 1 hour.

Prepare the broccoli, trimming the ends and removing any large leaves or thick stalks. Set aside.

Heat the oil in a heavy-based saucepan to 375°F (190°C) or until a cube of bread browns in 30 seconds. Dip the broccoli in the batter and fry in batches for 3–4 minutes or until golden and cooked. Remove from the oil with a slotted spoon and drain on paper towel.

Serve immediately with the dipping sauce.

■ *The dipping sauce will keep for one month stored in a sterilized jar.*

Tenderstem broccoli tempura with soy chili dipping sauce

Warm artichoke and caper dip

Warm artichoke and caper dip

Rich

My good friend Lynne Patterson is the real inspiration behind this dip.
Don't worry, Lynne, people don't invite you to potlucks just because of this! My
version is lightened up by substituting yogurt for mayonnaise.

Makes 14fl oz (400ml)

14-oz (400-g) **can or jar of grilled
    or ungrilled artichoke hearts,**
    drained and chopped
¼ cup (30g) **freshly grated Parmesan
    cheese**
Generous 1 cup (250g) **thick, full fat
    plain, natural yogurt**
1 **garlic clove,** peeled and minced
1 Tbsp **fresh lemon juice**
2 Tbsp **capers,** rinsed
Sea salt and pepper to taste
Crusty bread or toasted pita
    to serve

Preheat the oven to 350°F (180°C).
Combine the ingredients (except the
bread/pita) in a ovenproof serving dish
and bake for 10 minutes.

Serve warm from the oven with crusty
bread or toasted pita.

Béchamel sauce

Makes 2¼ cups (500ml)

2¼ cups (500ml) **milk**
2 **bay leaves**
¼ stick (30g) **butter**

¼ cup (30g) **all-purpose flour**
Pinch **freshly grated nutmeg**
Salt and freshly ground black pepper
　to taste

Put the milk and bay leaves into a saucepan and bring to a boil. Remove from the heat and set aside.

In a small heavy-based saucepan, melt the butter. Remove from the heat and stir in the flour using a small whisk. When the mixture is smooth, return to the heat and cook, stirring constantly, for a further 2–3 minutes or until the mixture darkens slightly in color and the flour is cooked. Remove from the heat again and gradually stir in the hot milk, whisking constantly, until all the milk is added and the mixture is smooth. Return to a low heat and cook for a further 10–15 minutes, stirring continuously, ensuring your whisk or spoon covers the complete surface of the bottom of the pan. Season to taste with the nutmeg, salt and pepper.

Béchamel sauce can be kept in the refrigerator for three days in an airtight container.

Egg pasta

Serves 4

1⅔ cups (200g) **all-purpose flour**
1 cup (100g) **semolina flour**

½ tsp **salt**
3 **eggs**
1 Tbsp **olive oil**

Place the flours and salt in a large bowl and mix together. Make a hollow reservoir in the center and crack the eggs into it and add the oil. With a fork slowly break up the eggs and draw the flour in to make a paste. Keep going until all of the flour has been mixed in and it forms a ball. If it is too damp add a little more flour, and if too dry add a little more water. Knead the dough until it is soft and silky, and when you press your finger into it the depression bounces out. This will take about 10 minutes (but depends on how consistently you knead). Wrap the pasta in clear food wrap and refrigerate for 30 minutes.

When it comes to rolling the pasta, you can either use a pasta machine, following the directions supplied, or alternatively roll the dough on a lightly-floured work-surface using a rolling pin. Roll the pasta, turning occasionally, until it is thin enough for you to see your fingers through it. Leave the pasta to dry in sheets or cut into shapes as desired.

Fig and red onion relish

Makes about 1 cup (200g)

2 Tbsp **extra-virgin olive oil**
1 **red onion,** peeled and finely sliced
1 tsp finely chopped **fresh rosemary**

⅔ cup (150ml) **red wine**
½ cup (100ml) **balsamic vinegar**
¼ cup (50g) **superfine sugar**
3 medium (150g) **plump dried figs,**
  quartered

Heat the oil in a nonstick frying pan or sauté pan. Add the onions and sauté until soft but not browned, about 6–8 minutes. Add the rosemary and sauté for a further minute before adding the wine and vinegar. Keep the heat at medium until the liquid has reduced by half. Add the sugar and figs and reduce the heat to low. Cook, stirring occasionally, for 15–20 minutes or until the mixture is the consistency of a syrupy jam. Remove from the heat and set aside to cool.

Moroccan tomato jam

Makes 1¾ cups (400g)

1 Tbsp **olive oil**
1 **onion,** peeled and finely chopped
Pinch **saffron strands**
¼ tsp **ground cinnamon**

½ tsp grated, peeled **fresh root ginger**
14-oz (400-g) can chopped tomatoes
2 Tbsp **honey**
Salt and freshly ground black pepper

Heat the oil in a medium saucepan and cook the onion for 3–4 minutes or until cooked and translucent but not browned. Add the saffron, cinnamon and ginger and cook for a further minute. Add the tomatoes and honey. Bring to a boil then reduce the heat and simmer for about 30 minutes or until very thick and jam-like. Season with salt and pepper then leave to cool to room temperature.

Spicy mango salsa

Makes 2¼ cups (500ml)

1 **large mango,** peeled and diced
1 **fresh tomato,** finely diced
2 Tbsp **red wine vinegar**
3 Tbsp **red onion,** finely chopped
4 **green onions, white and green
parts only,** thinly sliced
2 Tbsp **flat-leaf parsley or cilantro,**
finely chopped

½ tsp **flaky sea salt**
Juice of 1 **lime**
1 **clove garlic,** minced
1 Tbsp **pickled jalapeno chili juice**
1 Tbsp **pickled jalapeno chili,**
chopped
½ **red pepper,** finely diced
Pinch **ground cumin**
Pinch **cayenne**

Combine all ingredients in a glass bowl. Mix well, then refrigerate to allow flavors to
infuse, about 1 hour. Serve at room temperature.

Rhubarb chutney

Makes 1 cup (250ml)

1 tsp **mustard seeds**
7oz (200g) **rhubarb,** chopped into
   ½-in (2-cm) slices
Zest of 1 lemon
1 **small red onion,** peeled and finely
   chopped
1 Tbsp finely chopped, peeled **fresh
   root ginger**

¼ cup (75ml) **cold water**
⅓ cup (50g) **dried cranberries**
⅓ cup (50g) **raisins**
3 Tbsp **brown sugar**
2 Tbsp **balsamic vinegar**
**Sea salt and freshly ground black
   pepper to taste**

Heat a saucepan on medium heat. Add the mustard seeds and toast until they begin to pop. Add the remaining ingredients and simmer until the rhubarb is tender, about 5 minutes. Turn up the heat and bring to a boil. Boil gently for a further 5 minutes, until thickened. Season with salt and pepper to taste.

Serve immediately or store in a sterilized, airtight jar for up to two weeks.

Salsa verde

Makes ⅔ cup (150g) approximately

2 cups (40g) **fresh flat-leaf parsley**
¼ cup (5g) **fresh mint leaves**
¼ cup (5g) **fresh basil leaves**
6 (25g) **anchovy fillets**
1 **garlic clove,** peeled and roughly chopped

1 Tbsp **capers**
¼–½ cup (75–100ml) **extra-virgin olive oil**
Sea salt and freshly ground black pepper

Combine all the ingredients, except the olive oil and seasoning, in a food processor and blend to a rough paste. Drizzle in the oil, with the motor running, until the desired texture is reached. Depending on its use you may prefer a denser or thinner end product.

Sambal kacang

Makes 1¾ cups (400ml)

1 Tbsp **vegetable, seed or peanut oil**
4 **shallots** (approx. 5oz/150g), peeled and chopped
2 **garlic cloves,** peeled and chopped
½ tsp **cayenne,** or to taste

½ tsp **chili powder,** or to taste
Scant ½ cup (200ml) **carrot juice**
1 Tbsp **dark soy sauce**
1 tsp **cider vinegar**
1 cup (250ml) **natural, chunky peanut butter**

Heat the oil in a large saucepan over medium heat. Add the shallots and garlic and sauté until soft. Add the spices and stir until fragrant, cooking for 1 minute. Add the remaining ingredients and stir well – it will be quite thick – over medium heat. Remove from the heat until ready to serve.

Before serving, reheat and add water, a few Tbsp at a time, to thin the sauce. Taste the sauce; it might need a little more soy sauce or if heat is what you're after, a pinch of cayenne.

The sauce will keep in a sealed container in the fridge for up to one week.

Sweet Thai chili sauce

Makes 1½ cups (350ml)

4 **garlic cloves,** peeled and roughly chopped
2 **large red chili peppers,** stems removed
2 tsp peeled, grated **fresh root ginger**
Grated rind of 2 **limes**

2 **stalks lemongrass,** roughly chopped
½ cup (10g) **fresh cilantro leaves and stalks**
1 cup (200g) **superfine sugar**
6 Tbsp (75ml) **cider vinegar**
¼ cup (75ml) **fish sauce**
3 Tbsp **light soy sauce**

Purée the first six ingredients to a paste in a food processor.

Put the sugar in a medium-sized saucepan with 3 Tbsp of cold water and heat, stirring, until the sugar dissolves. Remove the spoon, increase the heat and boil gently for 5–6 minutes, until light golden caramel in color. Carefully stir in the paste. Bring back to a boil and add the vinegar, fish sauce and soy sauce. Bring back to a boil again and simmer for a further minute. Remove from the heat and leave to cool before serving.

The sauce will keep for up to one month, stored in a sterilized, airtight jar in the fridge.

Thai chili paste

Makes 4–5 Tbsp

3–4 **birdseye chilies,** halved and deseeded
2 **shallots,** peeled and finely chopped
2 **garlic cloves,** peeled and roughly chopped
¾-in (2-cm) **piece fresh root ginger,** peeled and roughly chopped

3 **kaffir lime leaves,** central vein removed and finely shredded
2 tsp **coriander seeds,** toasted
1 tsp **cumin seeds,** toasted
2–3 **cardamom pods**
2 Tbsp **vegetable oil**
1 tsp **fish sauce**
1 large handful **fresh cilantro leaves and stalks**

Chop the dried chilies into ½-in (1-cm) pieces and soak in warm (boiled) water until softened, about 20 minutes. Drain well.

Heat a small frying pan over high heat. Add the coriander seeds and toast until they pop and are fragrant, 2–3 minutes. Remove from the heat and cool. Grind in a mortar with a pestle.

Combine all of the remaining ingredients, except for the soy sauce, in a food processor and pulse until combined. Add enough soy sauce to make a slightly runny paste.

The paste is best the day it's made, but if you cover the surface directly with clear food wrap it will keep in the fridge for up to one week.

Root vegetable and tamarind gravy

Makes 2¼–2½ cups (500–570ml)

3 Tbsp **olive oil**
1 **large red onion,** peeled and roughly chopped
1 **turnip,** peeled and roughly chopped
2 **celery stalks,** roughly chopped
2 **carrots,** peeled and roughly chopped
2½ cups (200g) **mushrooms,** roughly chopped

2 **garlic cloves,** peeled
6¼ cups (1.5L) **vegetable stock**
1 **sprig fresh rosemary**
1 **sprig fresh thyme**
1 Tbsp **tamarind**
4 Tbsp **balsamic vinegar**
4 Tbsp **soy sauce**
½ cup (50g) **dates,** chopped
¼ tsp **crushed red pepper**
Sea salt and ground black pepper to taste

Heat the oil in a large stockpot over medium-high heat. Cook the vegetables and garlic, stirring occasionally until nicely browned, approximately 20 minutes.

Add the remaining ingredients, except for the salt and pepper, and bring to a boil. Cover and simmer for 30–40 minutes, or until turnip and carrots are very tender.

Strain the ingredients into a clean pot through a large sieve. Ease the pulp through with the back of a wooden spoon. Eat or discard remaining vegetables.

Season the gravy with salt and pepper to taste.

Serve or freeze in an airtight container.

Vegetable stock

Makes 7½ cups (1.75L)

2 **celery stalks with leaves,** roughly chopped
2 **leeks,** roughly chopped
1 **onion,** unpeeled, halved
2 **carrots,** unpeeled, roughly chopped

1 **head of garlic,** cut in half horizontally
5 **sprigs fresh parsley**
3 **bay leaves**
5 **sprigs fresh thyme**
1 tsp **black peppercorns**
9 cups (2L) **cold water**

Put all the ingredients in a large stockpot and cover with the water, adding extra if needed, to ensure everything is covered by about 1in (2.5cm). Bring to a gentle simmer, removing any scum with a large spoon, and simmer for about 1 hour. Leave to cool slightly before straining through a fine sieve.

Glossary

Adobo sauce – A Mexican, dark red sauce made from ground chilies, herbs and vinegar. Delicious as a marinade or served as a sauce. Chipotle chilies are often packed in adobo sauce.

Aioli – Originally from Provence, a garlic mayonnaise made with garlic, egg yolks and olive oil. The Spanish have a similar garlic-flavored mayonnaise that is called *alioli*.

Chèvre – A cheese made from goat's milk.

Chipotle chili – Thick-fleshed jalapeño chilies that have been smoked and dried. The result is a rich, flavorful, moderately hot chili.

Coconut milk – Made by combining water and shredded coconut, which is then simmered and strained through muslin.

Cilantro – Also known as coriander or Chinese parsley, cilantro is a fragrant, parsley-like plant popular in Thai, Indian and Latin American cooking. It is one of the only plants used as both a herb and a spice.

Coriander seeds – Quite different in taste to cilantro leaves, they are a mild, golden-brown, dry-roasted seed used in soups, stews, curry powders, pickling brines and marinades.

Crème fraîche – A thickened cream – thicker than sour cream – with a tangy, nutty flavor. Crème fraîche is an ideal topping for soups, as it won't curdle if boiled.

Crudités – Raw or blanched vegetables that are served cut into strips or broken into florets and eaten using your fingers. Usually served as a snack or *hors d'oeuvre*.

Dukkah – Of Egyptian origin, a blend of nuts, seeds and spices.

Fish sauce – A Southeast Asian condiment made from salted, fermented fish, used to salt dishes. The Thai version is called *nam pla*, while the Vietnamese call it *nuoc nam*.

Jalapeño – Bright green, medium-hot chili with thick, crisp flesh.

Jerusalem artichokes – Also known as a "sun chokes," this knobby, brown tuber is a member of the sunflower family. They are available in late autumn through early winter. Their skin is edible and highly nutritious, so scrub well instead of peeling if possible. Jerusalem artichokes can be eaten raw or cooked.

Kaffir lime leaves – Sometimes sold fresh, but most often frozen, these dark green leaves have an unmistakable Thai-like, citronella flavor. Leaves will keep in an airtight bag in the freezer for several months.

Kelp – The general name for any edible seaweed.

Marsala – Deep amber-colored, aromatic wine produced in the Marsala region of Sicily.

Miso paste – A staple in Japanese cooking. It is fermented soybean paste that looks much like peanut butter. There are three types of miso paste available: barley miso, rice miso and soybean miso. All are injected with a mold, then aged from six months to three years. Flavor depends on the amount of mold injected, the amount of salt used, and the period of fermentation.

Palm sugar – This dark, coarse, unrefined sugar, also known as "jaggery," can be made either from the sap of palm trees or from sugar-cane juice. It is primarily used in Southeast Asia and comes in several forms, the two most popular being a soft, honey-butter texture and a solid cake-like form.

Parmesan cheese – A dry, hard cow's milk cheese with a sharp flavor. Parmesan cheeses are made throughout the world, but the best is Italy's Parmigiano Reggiano. It has a grainy, melt-in-your-mouth quality that comes from at least two years of ageing. Look for the words "Parmigiano Reggiano" stenciled on the rind, which means the cheese was produced in and around Parma, where the cheese originated.

Puy lentils – Lentils are the easiest of all légumes because they don't require pre-soaking before cooking. There are many varieties, but we like French puy lentils best for their flavor and good-temperedness. They are tiny, slate-gray-green in color, and retain their shape and texture when cooked.

Sea salt – The salt produced after the evaporation of sea water. It is sold as fine grains or large, flaky crystals. Sprinkled on a soup with a few twists of black pepper, it adds flavor, texture and decadence.

Mascarpone – Soft, unripened cheese that belongs to the cream cheese family. It comes from Switzerland and Italy and is a thick, rich, sweet and velvety, ivory-colored cheese produced from cow's milk that has the texture of sour cream. It is sold in plastic tubs and can be found in speciality food stores and in the deli section of most supermarkets.

Okra – Slender, five-sided pod containing white seeds. When cooked, they release a sticky substance that is used to thicken stews, soups and curries.

Orzo (riso, pasterelle) – Small pasta shape that looks similar to rice.

Polenta – Made from cornmeal and eaten prolifically in northern Italy. It is graded depending on its coarseness. Polenta can be eaten soft or spread onto trays, cooled until set and then fried.

Quinoa – This ancient grain is experiencing a renaissance. Quinoa contains more protein than any other grain, giving it the reputation as the "supergrain of the future." The tiny, swirly grains cook like rice, but cook in half the time. The flavor is delicate, much like couscous.

Ricotta – The word ricotta literally means "recooked" in Italian. It is made from the whey of other cheeses that is cooked again to make a milky-white, soft, granular and mild-tasting cheese. Ricotta does not melt.

Sea vegetables – A staple in Japanese diets for centuries, often referred to as seaweed. There are numerous varieties of sea vegetables that come in all shapes and colors. Sea vegetables are neither plants nor animals, but are classified in a group known as algae.

Sumac – Middle Eastern spice used extensively in Iran, Turkey and Lebanon. It is made from a reddish berry and has a slight astringent, lemony flavor.

Tagine – Dish from North Africa that is characterized by its sweet and savory mix of flavors. It is sometimes translated to be a "stew."

Taleggio – Semi-soft cheese made from whole cow's milk. Its flavor can range from mild to pungent, depending on its age. When young, taleggio's color is pale yellow. As it ages it darkens to deep yellow and becomes rather runny. Taleggio is sold in flat blocks or cylinders and is covered either with a wax coating or a thin mold.

Tofu – Cheese-like curd made from soybeans. It is an excellent source of protein and low in fat. Although bland in flavor, tofu easily absorbs the flavor of the ingredients it is cooked in. Available in different varieties e.g. soft (silken), firm, deep-fried and sheets.

About the authors

After seven years of living and working in London, Pippa Cuthbert has returned to her homeland, New Zealand, to start a family and get her vegetable garden under way. Pippa continues to work as a food writer and stylist on books, magazines, packaging, advertising and TV commercials.

Food and writing are Lindsay Cameron Wilson's passions, so she blended the two in college where she studied history, journalism and the culinary arts. She has since worked in the test kitchens of *Canadian Living Magazine* in Toronto and *Sunset Magazine* in San Francisco. In 2001 she left her job as a food columnist in Halifax, Nova Scotia, and moved to London. That's when she met Pippa, and the work for their first book *Juice!* began. Fueled by juice, the two moved on to *Ice Cream!*, *Soup!*, *Grill!*, *Pizza!*, *Cookies!* and now *Vegetables!* Lindsay continues to work as a food journalist in Canada, where she now lives with her husband James, and sons Luke and Charlie.

Index